The accidental universe

The accidental universe

P.C.W. DAVIES

Professor of Theoretical Physics, University of Newcastle-upon-Tyne

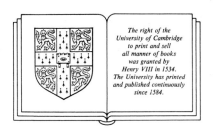

The right of the
University of Cambridge
to print and sell
all manner of books
was granted by
Henry VIII in 1534.
The University has printed
and published continuously
since 1584.

CAMBRIDGE UNIVERSITY PRESS

Cambridge

London   New York   New Rochelle

Melbourne   Sydney

Published by the Press Syndicate of the University of Cambridge
The Pitt Building, Trumpington Street, Cambridge CB2 1RP
32 East 57th Street, New York, NY 10022, USA
10 Stamford Road, Oakleigh, Melbourne 3166, Australia

First published 1982
Reprinted 1983, 1984, 1985

Printed in Great Britain at the University Press, Cambridge

Library of Congress catalogue card number: 81-21592

British Library cataloguing in publication data
Davies, P.C.W.
The accidental universe.
1. Cosmology
I. Title
523.1'1   QB981
ISBN 0 521 242126 hard covers
ISBN 0 521 286921 paperback

CONTENTS

In spite of the spectacular progress made by physicists in recent years in understanding the basic forces of nature, many fundamental features of the physical world seem to be arbitrary and meaningless. Why are there three space dimensions? Why is gravity so weak? Why is the proton 1836 times heavier than the electron? And so on.

The numerical values that nature has assigned to the fundamental constants, such as the charge on the electron, the mass of the proton, and the Newtonian gravitational constant, may be mysterious, but they are crucially relevant to the structure of the universe that we perceive. As more and more physical systems, from nuclei to galaxies, have become better understood, scientists have begun to realize that many characteristics of these systems are remarkably sensitive to the precise values of the fundamental constants. Had nature opted for a slightly different set of numbers, the world would be a very different place. Probably we would not be here to see it.

More intriguing still, certain crucial structures, such as solar-type stars, depend for their characteristic features on wildly improbable numerical accidents that combine together fundamental constants from distinct branches of physics. And when one goes on to study cosmology – the overall structure and evolution of the universe – incredulity mounts. Recent discoveries about the primeval cosmos oblige us to accept that the expanding universe has been set up in its motion with a cooperation of astonishing precision.

Many of these 'accidents of nature' have been known for decades. In the 1930s, Eddington and Dirac were struck by the curious and unexpected concurrence of certain very large numbers computed from atomic physics and cosmology –

apparently unrelated topics. These, and other examples, give the impression of a universe that is delicately balanced in a variety of ways.

The only systematic attempt (outside religion) to explain the extraordinarily contrived appearance of the physical world has developed out of a radical departure from traditional scientific thinking. Called the *anthropic principle*, the idea is to relate basic world features to our own existence as observers. The principle has its origins with great physicists such as Boltzmann, and in recent years has been restated by a number of eminent scientists, including Brandon Carter, Robert Dicke, Freeman Dyson, Stephen Hawking, Martin Rees and John Wheeler. Some of these scientists go so far as to claim that our existence can be used as a biological selection effect, allowing one to actually explain the otherwise mysterious numerical values of the fundamental physical constants.

Although some writers have found the philosophical basis of the anthropic principle objectionable, it is difficult not to be struck by some of the surprisingly fortuitous accidents without which our existence would be impossible. This book surveys some of these accidents and numerical coincidences, and only in the final chapter is the issue of the anthropic principle raised.

Intended for the general reader, the treatment is non-specialist, and will appeal to both scientists and scientifically-inclined laymen alike. Students of philosophy and science will find the text easy to follow in most parts, and will require only a general familiarity with basic physics. Chapter 1 summarizes much of the physics needed for the subsequent chapters. The level corresponds roughly to that of *Scientific American* or *New Scientist*. Where mathematics is used, it almost always involves only elementary algebra.

Much of the treatment presented here follows the pattern of some excellent technical surveys already published. Rather than interrupt the text with references, I have instead given each chapter a bibliography.

I am especially indebted to Dr Bernard Carr and Professor Martin Rees, on whose review article much of this book is based. I have received many helpful comments and suggestions from these authors, as well as from Dr John Barrow, Dr Frank Tipler and Dr John Leslie. I have also benefited from several useful discussions with members of the Physics and Philosophy departments of the University of Canterbury, New Zealand.

P.C.W. Davies

NOTE ON UNITS AND NOMENCLATURE

The mathematical relations of most interest in this book are not exact equations, but inequalities or approximate equalities. The symbol $\sim$ is most frequently employed, and means that two quantities are equal to within an order of magnitude or so. For example, $7 \times 10^8 \sim 5 \times 10^9$. Where used in front of a single symbol, $\sim$ means 'of this order': for example $\sim 10^3$ means a number, such as 630 or 2018, that is of the same order of magnitude as 10^3. The symbols $>$ and $<$ mean 'greater than' and 'less than', as usual, while $\gtrsim$ means 'greater than about' a certain number, with a corresponding meaning for $\lesssim$.

Sometimes the symbol $\simeq$ is used. This is the approximate equality sign and is applied when two quantities are equal to within a factor of two or so. Thus $\pi^2 \simeq 10$. Finally $\equiv$ is used to mean 'defined by': for example $\alpha \equiv e^2/4\pi\varepsilon_0\hbar c$ means that α is the shorthand symbol for the quantity $e^2/4\pi\varepsilon_0\hbar c$. Readers who wish to rework the expressions to greater accuracy will find a table of numerical values for the fundamental constants on page 39, and further useful data on page 79.

Throughout the text SI units are used. The reader should be warned that almost all the references cited in the bibliography use either c.g.s. units (especially astronomical works) or special units where all, or some, of the constants $\hbar$, c, G and k are put equal to 1. Because the SI convention for electric charge in free space requires that e^2 is accompanied by $(4\pi\varepsilon_0)^{-1}$, where ε_0 is the permittivity (dielectric constant) of free space, the 4π factors associated with this will be carried through the expressions explicitly, even when other numerical factors and powers of π have been removed as part of the $\sim$

approximation. Moreover, although the style ε_0 is conventional, we shall never have occasion to consider dielectric media, so for compactness the zero subscript on ε will be suppressed.

1

The fundamental ingredients of nature

The variety and complexity of physical systems that adorn our universe are so bewildering that the task of discovering simple laws to describe them all appears hopeless. Yet remarkable though it may seem, the fundamental principles that control objects as diverse as atoms and stars are well enough understood that an integrated account can be given of most of the more common systems in the natural world. Our ability to summarize the workings of nature within a single theoretical framework stems from the fact that the really fundamental features of physics are both simple and comprehensive. Theories like quantum mechanics have such enormous predictive power that they explain at a stroke phenomena as dissimilar as the formation of a crystal and the collapse of a neutron star.

The universality of fundamental physics lies behind the account given in the forthcoming chapters. The reader will discover that although the specific details of physical systems can only be determined from complicated analyses, the broad structural features are largely determined from a few elementary considerations. These considerations reveal a universe full of stunning surprises.

1.1 Structure on all scales

Nature displays a hierarchy of structure. From the smallest known constituents of the atom to the large scale arrangement of the galaxies, we observe systems with characteristic organization and size, each level of structure interlocking with the others in a highly ordered way. What determines the scale of these structures and their relation to one another? Why are galaxies so big and atoms so small? Why are stars so hot and the night sky so dark?

1

The largest familiar structure is the galaxy, of which our own Milky Way is typical. Containing about 10^{11} stars it is shaped like a plate with a central ball of more densely packed stars. The whole assemblage of stars, together with clouds of gas and some dust grains, slowly rotates. The stars are not distributed uniformly throughout the galaxy, but tend to concentrate in spiral shaped arms. A typical galaxy is about 10^5 light years in diameter.

Galaxies tend to cluster together throughout space in groups ranging from a few dozen to many thousands. There is good observational evidence that above this scale of structure the universe is remarkably uniform in the way that matter and radiation are distributed. The uniformity is both in

Fig. 1. The Virgo cluster of galaxies is one of the nearest clusters to our own Group. (Scale of print 26 arcsec/mm.)
Courtesy of the UK Schmidt Telescope Unit, Royal Observatory, Edinburgh

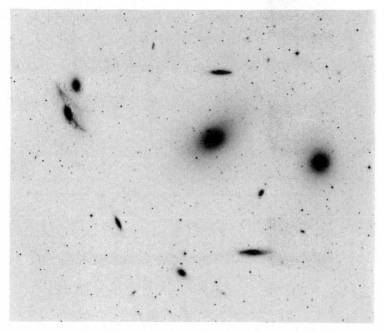

orientation about us (isotropy) and from region to region in distance from us (homogeneity).

The entire assemblage of clusters of galaxies is not, of course, at rest. The force of gravity is forever trying to coalesce the dispersed material into more compact clumps, so that all matter is engaged in a struggle between gravity and the opposing forces of dispersal. In relatively small objects, such as stars and planets, gravity has partially won the struggle. The density of material in these objects is about 10^{30} times higher than the average cosmic density of matter.

The larger systems – galaxies and clusters of galaxies – have

Fig. 2. A typical spiral galaxy, in the constellation Triangulum. Our own Milky Way galaxy would have a similar appearance from afar.

avoided collapse because they are rotating, and orbiting around each other. Gravitational implosion is counterbalanced by centrifugal effects. In addition to this, the clusters of galaxies are prevented from falling together by the fact that the entire universe is engaged in a systematic pattern of expansion, each cluster gradually moving away from its neighbours. The expansion of the universe, discovered by Edwin Hubble in the 1920s, is a cornerstone of modern cosmology, and is best envisaged as the continual swelling or stretching of space itself. As the space between galaxies expands, so the galaxies grow further apart.

As with the distribution of matter, this expansion is unexpectedly uniform throughout the universe. Because the universe on the very large scale is so uniform in its arrangement, the motion of the whole assembly can be characterized by a single parameter: the rate at which two typical galaxies a certain distance apart are separating. This is called the Hubble constant, denoted H, and its value is often quoted by astronomers as about 50 kilometers per second per megaparsec, which means that two galaxies, say, 10 megaparsecs (about 30 million light years) apart, are receding from each other at about 500 kilometers per second. In more familiar units $H \simeq 10^{-18}$ s^{-1}.

If the galaxies are currently moving apart then they must have been closer together in the past. The units of H are those of velocity/distance, which is inverse time; inverting H therefore yields a fundamental unit of time – the Hubble time – by which to gauge cosmological change. The value of H^{-1} is about 10^{10} years, which implies that 10^{10} years ago the large scale structure of the universe must have been very different from today, with the galaxies crowded much closer together.

As the universe slowly expands, so the intergalactic gravitational forces operate to restrain the dispersal of the galaxies. One would therefore expect the expansion rate H to be gradually slowing down, much as a vertically propelled projectile gradually decelerates. There is indeed some observational evidence that the cosmological expansion rate is diminishing.

If one accepts the idea of a decelerating cosmos, then it follows that 10^{10} years ago the expansion rate was higher than now. Going backwards in time, one expects an accelerating rate of expansion in order for the galaxies to escape falling together under their mutual gravity. Extrapolating as far back as one can, it seems that about 18 billion years ago the universe was infinitely compressed, and expanding infinitely rapidly. This dense, explosive phase is popularly called the big bang, and because it began a finite time ago it is generally held to describe the actual creation of the universe.

Fig. 3 shows how a typical volume of space (for example a cubic light year as measured today) has expanded from nothing since the big bang. Notice the rapid deceleration of the expansion rate in the early stages, followed by the steady decline, expected to continue in the future. This diminishing deceleration is due to the fact that gravity weakens with separation. As the galaxies become more dispersed so the

Fig. 3. The expanding universe. Space is continually swelling, thereby diluting the density of matter and sweeping the galaxies apart from each other. The curve shows how the diameter of a typical spherical volume of space grows at a diminishing rate. The present expansion rate, H, defined as $\dot{a}/a$, where a is the radius of some typical spherical volume of space, is given by the tangent to the curve at the instant marked 'now'. It yields a characteristic time, H^{-1}, called the Hubble time, which is about one and a half times the age of the universe.

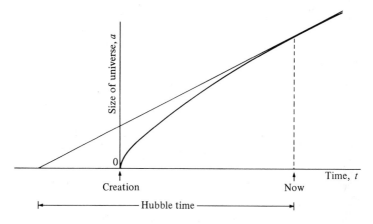

intergalactic gravitational forces diminish their restraint on the expansion. Because of the early rapid deceleration in the expansion rate, the Hubble time, H^{-1}, is roughly the same (to within a factor of about $\frac{3}{2}$) as the age of the universe. Note that H is therefore not actually a constant, in spite of its name.

Although the galaxies appear to be islands of glowing matter surrounded by vast chasms of empty space, the intergalactic regions are not totally void. There is undoubtedly some tenuous dark or transparent matter there. More important, all of space, including these superficially empty chasms, is filled with heat radiation. This radiation bathes the entire universe in a rather feeble glow – the temperature is about 3 K. It is the extreme isotropy of this cosmic background heat radiation as received on Earth that provides such good evidence of the large scale uniformity of the universe, for the heat radiation has travelled unimpeded over cosmological distances, and would have carried the imprint of any large scale irregularities.

One of the fundamental mysteries of modern cosmology is why the temperature of the cosmic heat radiation is 3 K rather than some other value. In fact, as the universe expands, the temperature falls. However, the ratio of the number of thermal photons to the number of, say, protons or electrons, in some large volume of space, is unchanged by the cosmic expansion (this will be shown in detail in Section 2.4). The photon/proton ratio is denoted by S, and has a value of about 10^9. Evidently, photons are considerably more numerous than atoms.

Turning now to length scales smaller than galaxies one identifies the most familiar structures in the universe: stars. Stars are held in equilibrium by the balance of their own gravitational force, which tries to shrink them, with thermally generated internal pressure sustained by nuclear reactions in the interior. The smaller, cooler planets overcome their self-gravity by solid state forces which are basically electric in origin. Stars are frequently found in clusters of up to a million in number.

Reducing in scale still further, one encounters large living organisms (including man), representing the most developed structures, in terms of complexity, yet known. Passing on down in size through cells and biologically active chain molecules such as DNA, one reaches the level of atoms, now known to be composite systems with their own internal structure.

The nuclei of atoms consist of two types of particles: electrically charged protons, and neutrons. Both have a mass of about 10^{-27} kg. In isolation, neutrons decay, with an average lifetime of a few minutes, into protons and electrons. In addition, another particle is emitted, called an antineutrino (the antiparticle of a neutrino – see Section 1.3). Neutrinos are electrically neutral, possess little or no mass, and interact so weakly with ordinary matter that they easily pass right through the Earth. Neutrinos are therefore extremely elusive, and only since the Second World War has their existence been unequivocally confirmed. They nevertheless play an important role in the structure of the universe.

The proton is the fundamental building block of nuclear structure (see Fig. 4). The chemical elements are determined by the numbers of protons contained in the nucleus. The nucleus of the simplest element, hydrogen, consists of a single proton. An isotope of hydrogen, called deuterium, has a neutron and proton stuck together. The nuclear charge is therefore the same as for ordinary hydrogen, but the nucleus is roughly twice as massive.

The next simplest element is helium, which in its normal form contains two protons and two neutrons. Continuing upwards, lithium has three protons, beryllium four, etc. Important elements are carbon, with six protons, oxygen with eight, iron with 26 and uranium with 92. Heavy elements, such as uranium, usually contain about one and a half times as many neutrons as protons. Many are radioactive. Elements heavier than uranium decay with average lifetimes less than the age of the Earth, so are not found in abundance on Earth.

Being a particle of fundamental importance, the proton's properties are of special significance for nuclear (and ultimately atomic and chemical) physics. These properties

Fig. 4. The chemical elements. The chemistry of an atom is determined by the nuclear charge (numbers of protons) which in normal form is exactly balanced by the numbers of electrons. When an atom loses electrons it is described as being *ionized*. The outer electrons help form the bonds that stick atoms together into molecules. The heaviest, most complex atoms contain about 250 nuclear particles and about 90 electrons.

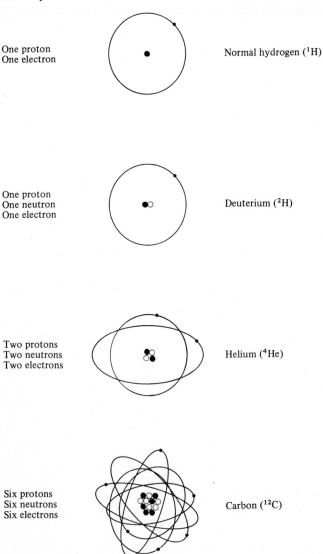

One proton
One electron   Normal hydrogen (^{1}H)

One proton
One neutron   Deuterium (^{2}H)
One electron

Two protons
Two neutrons   Helium (^{4}He)
Two electrons

Six protons
Six neutrons   Carbon (^{12}C)
Six electrons

include its mass, electric charge and size. The size of a proton is a rather subtle concept which will be discussed in Section 1.3. For now we take it to be about 10^{-15} m. This distance can be converted to a fundamental unit of time: the duration for light to cross the proton. The speed of light is the fastest rate at which information can travel and is therefore of special significance. The light travel time across a proton is about 10^{-24} s. Physically this is the smallest interval required for a proton to behave in an integrated way as a single entity.

We therefore now have two natural time scales: the Hubble time, $t_H \sim 10^{10}$ years, and the nuclear time, denoted $t_N \sim 10^{-24}$ s. Their ratio is the impressively large number 10^{41}. The origin of this number, and why it is so big, will be an important topic in the coming chapters.

1.2   The forces of nature

As far as we know, all the various natural phenomena are controlled by just four fundamental forces: gravity, electromagnetism, and two nuclear forces called weak and strong. In recent years attempts have been made to describe these forces by a single mathematical theory. This so-called unification programme has interwoven the weak nuclear force with the electromagnetic force, and more recently made progress in incorporating the strong nuclear force too (see Table 1).

Gravity is familiar in daily life. It acts universally between all material bodies in the universe, and to good approximation (in the case of stars and planets) declines with distance in accordance with Isaac Newton's famous inverse square law. For two idealized point masses, each experiences a force directly towards the other mass of strength

$$F_{\text{grav}} = -\frac{Gm_1 m_2}{r^2}. \tag{1.1}$$

In this formula the negative sign indicates a force of attraction, r is the separation of the two bodies (the size of the bodies is assumed to be small relative to r), m_1 and m_2 are their respective masses. The constant G is universal and has an important significance. It controls the strength of the gravitational forces exerted by the masses. If m_1 and m_2 are taken to be some standard mass, say 1 kg each, and r a standard distance, say 1 m, then the observed force of attraction is 6.7×10^{-11} N. If G were larger, then this force would be larger in proportion. The assertion that G is a universal constant is the claim that, wherever in the universe, or at whatever point in history, one were to measure the force between two 1 kg masses at 1 m separation, then the result would always be 6.7×10^{-11} N. The quantity G must therefore be set alongside the other fundamental quantities as an important constant of nature that determines the structure of gravitating systems.

This century, Newton's theory of gravity has been replaced by a new theory, called the general theory of relativity, due to Albert Einstein. Although the results of general relativity differ somewhat from those of Newton's theory in the case of strong gravitational fields, the two theories coincide in the weak field limit, far from the gravitating bodies. Thus Newton's inverse square law, and the significance of the constant G, remain valid in Einstein's theory.

Table 1. *The forces of nature*

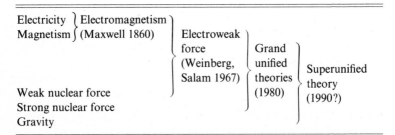

Ever closer scrutiny reveals that all of nature's disparate forces are really manifestations of a very small number – perhaps one – of fundamental forces. All known interactions can ultimately be reduced to just four basic types: electromagnetic, gravitational and two nuclear. The electromagnetic and weak nuclear forces, while physically very different in their operation, are actually two aspects of a single, unified, electroweak force. Recent advances suggest that the strong nuclear force, very different again in character, can also be brought in to this scheme in a grand unified theory (GUT) Gravity alone remains to be incorporated.

However, the general theory of relativity does contain one feature over and above Newton's theory. It is possible to include an additional force of gravity alongside the usual inverse square attraction. This extra gravitational force is distinguished by two unusual properties. First, it is repulsive, and operates to push matter apart, rather than pull it together as is the case with Newtonian gravity. Second, the strength of the repulsion accumulates with distance, whereas 'ordinary' gravity diminishes in accordance with (1.1). Hence the effects of the additional force only become important on the very large scale. For that reason, Einstein called this new contribution the cosmological term. He originally introduced it in order to explain how the entire universe could avoid falling together under the mutual gravitational forces of attraction. He sought to balance the attraction of ordinary gravity with the new cosmological repulsion to achieve a static universe.

After the cosmological term was introduced into general relativity in 1917, it was discovered by Hubble that the universe is not in any case static, but in a state of expansion. The galaxies avoid collapsing on to each other by virtue of their recessional motion. Realizing this, Einstein regarded his cosmological term as a great mistake, and hastily abandoned it. In spite of this, there is no *a priori* reason to rule it out and, as we shall see, modern quantum field theory definitely requires that such a term be present. The strength of the repulsion, however, is as yet immeasurably small. If we write the repulsive force as

$$F_{\text{cosmic}} = \Lambda rmc^2, \tag{1.2}$$

where m is the mass of the repelled object, r its distance from the repelling body, c is the speed of light, then Λ is a constant with the units m^{-2}. (Notice that the force is independent of the repelling mass.) Current observations place the upper limit on Λ of 10^{-53} m^{-2}. Thus, two one kilogram masses one metre apart feel a force of attraction which is at least 10^{25} times larger than the cosmic repulsion. On the other hand, two galaxies of mass, say, 10^{41} kg separated by a distance of 10^6 light years (about 10^{22} m) would experience comparable attractive and repulsive forces if Λ were actually near to its upper limit. Thus, although Λ has not been measured we shall still take it to be a universal constant of fundamental significance for the large scale structure of the universe. Later it will be argued that Λ is not strictly constant.

Turning to electromagnetism, both electricity and magnetism owe their origin to electric charge. The force of interaction between two electric charges depends in a complicated way both upon their position and motion, with electric and magnetic effects interwoven. There is one simple case, however, which occurs when the two charges are at rest. If the magnitudes of the charges are e_1 and e_2 respectively, and each is concentrated at a point, then the mutual force between them is purely electrical in nature, and is given by a formula almost identical to (1.1):

$$F_{\text{elec}} = \frac{e_1 e_2}{4\pi\varepsilon r^2}. \tag{1.3}$$

The electric force is along the line joining the two charges and is attractive or repulsive depending on the relative signs of e_1 and e_2. Once again ε is a universal constant that decides the strength of the electromagnetic effects. It has the value 8.85×10^{-12} F m^{-1}. Thus two charges of one coulomb placed at rest one metre apart will experience a force of 8.99×10^9 N.

It is found that electric charge is always attached to certain subatomic particles, of which the electron and proton are the most familiar. The numerical amount of charge on these particles is always the same; it is a universal constant of nature. In SI units the value is 1.6×10^{-19} C. By convention, the charge on the proton is taken to be positive. We denote it by e.

As far as magnetic forces are concerned, there is no experimental evidence for magnetic charge in the same sense as electric charge; magnetic forces are generated entirely by electric currents (moving electric charges). Nevertheless, some modern theories of unified forces predict the existence of magnetic charge. However, it is not necessary to introduce a new fundamental unit, because it was shown by Paul Dirac that the quantity of magnetic charge carried by a particle can only be a multiple of a basic unit that is entirely determined by the fundamental unit of electric charge e. Thus e determines both the strength of electricity and, if it exists, magneticity.

Turning to the two nuclear forces, consider the weak force first. This force is responsible for many nuclear processes, one of the more familiar being the transmutation of neutrons into protons. The weak force thus manifests itself more through changes in particle identity than particle motions. Nevertheless its strength can still be characterized by a universal constant, g_w, which determines the rate at which weakly induced transformations such as neutron decay proceed.

The strong nuclear force is considerably more complicated in nature than the other forces. In its grossest form the strong force is responsible for binding together the protons and neutrons of atomic nuclei. Without the strong force, the nuclei would explode as a result of the protons' electric repulsion. In this crude case it is possible to introduce a quantity g_s, analogous to electric charge, but considerably larger, as the label 'strong' suggests. The concept of g_s is only of limited value, however. For one reason, the strong force does not obey an inverse square law like (1.1) or (1.3). Instead it declines rapidly to zero outside an effective range of about 10^{-15} m. Secondly, the protons and neutrons, as we shall see, are composite bodies, so must themselves be internally bound by a very strong force. The interproton and interneutron force is really only a complex vestige of this internal force. Because our understanding of the internal structure of protons and neutrons is still tentative and rudimentary, the less fundamental, but simpler, concept of g_s will be employed here as a rough measure of the strength of the strong interaction.

1.3   Quantum theory and relativity

In addition to the forces of nature, the structure of our world is determined by the laws that govern how bodies move about under the influences of these forces. On the scale of daily experience these laws are adequately described by

Newton's mechanics. There are, however, three circumstances under which Newton's laws fail.

First, if the speed of the bodies involved approaches the speed of light, then the motions are distorted from the Newtonian prediction because of the effects of special relativity. The velocity of light, denoted c, acts as an absolute upper limit on the·speed of all material systems. We may take c as another fundamental universal quantity of great significance in determining the arrangement of the universe.

Second, if the gravitational field becomes intense, then not only must one abandon Newton's law of gravity, but also Newton's mechanics. According to the general theory of relativity, gravity is a manifestation of a spacetime distortion akin to the curvature of a surface, such as a sphere. Moving in this distorted background, a particle will behave in a way that is different from the Newtonian prediction.

Finally, when the size of the system concerned is comparable to that of an atom, Newton's laws fail again. It is necessary to use the quantum theory to describe the behaviour of submicroscopic particles and fields.

The central feature of the special theory of relativity is that the speed of light is measured to be the same by all observers, no matter how they are moving. Thus c is a universal constant of nature. This paradoxical situation can only be understood by supposing that time and space intervals are not separately invariant, but change from one reference frame to another, leading to the famous time dilation and length contraction effects.

Because light is always observed to travel at c it is clearly impossible for an observer to reach (or exceed) the speed of light. This speed acts as a sort of barrier for the propagation of any material body or any influence whatever. All physical disturbances are restricted to travel at c or slower. For this reason the mechanics of rapidly moving bodies possess some strange features. For example, the energy E and mass m_0 (measured at rest) of a body moving at speed v relative to some observer are related by the formula

$$E = m_0 c^2 / \sqrt{(1 - v^2/c^2)}. \tag{1.4}$$

As v approaches c, the energy of the body becomes large without limit, which implies that an infinite amount of energy must be supplied to the body to accelerate it to the speed of light. At the opposite limit, when $v = 0$, the body is at rest, but the energy does not vanish. Instead

$$E_{\text{rest}} = m_0 c^2. \tag{1.5}$$

This rest energy is due entirely to the mass m_0 of the body, and not to any effects of motion. Sometimes this is expressed by saying that energy and mass are equivalent, or that energy has mass, or that mass is energy.

As an illustration, consider the sun, which radiates about 10^{26} J s^{-1}. This energy loss is equivalent to a mass of 4 million tonnes, so each second the sun gets 4×10^6 tonnes lighter.

To take another example, the nucleus of oxygen contains eight protons and eight neutrons bound tightly together. The mass of an oxygen nucleus is 2.655×10^{-26} kg. On the other hand the mass of eight individual protons and eight individual neutrons is 2.678×10^{-26} kg. The missing 2.3×10^{-28} kg is due to the energy lost when the nuclear particles were assembled into a bound system.

Sometimes Eq. (1.4) is written

$$E = mc^2, \tag{1.6}$$

where $m = m_0(1 - v^2/c^2)^{-\frac{1}{2}}$ is called the *relativistic mass*. Its value depends on the velocity v. The quantity m_0 is then referred to as the *rest mass*, being the value of m at rest ($v = 0$).

If a particle travels at the speed of light (and this is obviously the case for photons – see below) then Eq. (1.4) indicates that, in order for E to be finite, $m_0 = 0$. In that case the right hand side of (1.4) reads $0/0$ which can be a finite quantity. In fact, the energy of a particle of light depends on the frequency of light, and is given by Eq. (1.7) below. Thus, if a particle has zero rest mass, it travels at the speed of light. Often physicists call such a particle 'massless'. This description refers to *rest* mass. Such a particle still has relativistic mass $m = E/c^2$.

We now turn to a brief description of the quantum theory.

Quantum effects involve a blurring of the particle concept, so that in many ways an electron, for example, behaves in a manner more usually associated with a wave. Thus, electrons can diffract around objects and form interference patterns. In other ways too their behaviour seems erratic; they may tunnel through barriers or bounce off insubstantial obstacles. Similarly, the wave concept can be modified at the submicroscopic level, so that light, for example, which is an electromagnetic wave, is emitted, absorbed and scattered in a fashion suggestive of small particles or corpuscles. These *photons* were the original 'packets' or 'quanta' that gave the theory its name.

The scale at which these wave–particle peculiarities characteristic of quantum effects become important is determined by Planck's constant h, which has the value 6.6×10^{-34} J s. More often one has to deal with the quantity $h/2\pi$, which is denoted $\hbar$. In the context of particle-like behaviour of waves, h arises as the ratio of energy to frequency of a photon. Thus

$$E = hv. \tag{1.7}$$

For wave-like aspects of particles, h is the product of momentum p and wavelength λ. So

$$p = h/\lambda. \tag{1.8}$$

Quantum theory must be taken into account when the mechanical quantities of interest are comparable to h. For example, an electron in orbit inside an atom has a kinetic energy of about 10^{-19} J and an orbital period of about 10^{-15} s. The product of energy and time for one period is thus $\sim 10^{-34}$ J s, that is, comparable to h. We conclude that quantum effects profoundly modify the behaviour of atomic electrons, which is, of course, true.

Because quantum theory, applied to the electromagnetic field, yields a description of the transport of energy and momentum through the field in the form of discrete photons, this quantized disturbance must be taken into account in processes where electric charges and currents act on each other. In the classical theory of electrostatics, the inverse square law, (1.3), is described in field language by saying that the charge e_1 creates an electric field of force around it, and the charge e_2 interacts with that field at a point a distance r away. It is the interaction between e_2 and the field that produces the force. If e_1 were disturbed in some way, the effect of this would be transmitted to e_2 through the field, and e_2 would respond accordingly. In the quantum theory the disturbance is considered to be transmitted through the field in the same way, but in the form of photons. When e_1 is moved, it emits photons which are subsequently absorbed by e_2, causing it to move also. The electromagnetic force is therefore described in terms of the *exchange* of field quanta, acting rather like messengers, between the sources. (To extend this description to the electrostatic force itself, in the absence of charge disturbance, it is necessary to invent an additional sort of photon which differs from those that cause the sensation of light.)

The description of the other forces of nature based on the exchange of intermediate field quanta has also been developed (see Fig. 5). For example, the gravitational force may be attributed to graviton exchange. The weak force involves the exchange of particles called intermediate vector bosons. In the most recent theory of the weak force, both electrically charged and neutral particles (called W and Z particles respectively) are necessary. Both W and Z particles are very massive (many times heavier than protons). The strong force, which is more complicated, will be discussed in greater detail in the next section.

When quantum and relativistic effects combine, a new phenomenon can occur: the creation and destruction of subatomic particles. The relation $E = m_0 c^2$ suggests that a particle of rest mass m_0 can be created if the energy $m_0 c^2$ is supplied somehow. This is confirmed by experiment. For example, an electron can be created from a very energetic photon (a gamma ray). To conserve electric charge, it is necessary that the newly created electron is accompanied by the appearance of a 'mirror' particle with opposite (positive) charge. This particle, called the positron, was discovered in

1932. It has the same mass as the electron. We shall henceforth denote the electron by e^- and the positron by e^+.

Other particles can similarly be created out of energy. Protons and neutrons (denoted by p and n respectively) appear accompanied by their respective mirror particles, denoted $\bar{p}$ and $\bar{n}$. It is found that every type of particle possesses a mirror particle in this way, referred to as its *antiparticle*; $\bar{p}$ is an antiproton, $\bar{n}$ is an antineutron and e^+, the positron, is an antielectron. Antiparticles are referred to generically as *antimatter*. The creation of particles usually involves the simultaneous appearance of a particle–antiparticle pair, for example e^+, e^-, so is often referred to as pair creation. In some cases (for example the photon) a particle is indistinguishable from its antiparticle, so that such particles can be created singly.

The energy for pair creation, $2m_0c^2$, can be supplied in

Fig. 5. Quantum description of forces. The basic mechanism whereby forces are transmitted between source particles is depicted schematically here. The straight lines show the paths of the particles that are the sources of the force. When particle 1 is disturbed it emits a temporarily created, or 'virtual' quantum of the force field (for example a photon of the electromagnetic field) which is subsequently absorbed by the second particle. Particle 2 then suffers a disturbance in response. In this way the particles can exert influences on each other from a distance. A description of this sort applies to all the fundamental forces of nature.

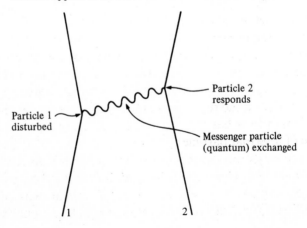

Particle 1 disturbed

Particle 2 responds

Messenger particle (quantum) exchanged

1   2

many different ways; for example, kinetic energy liberated on the impact of two other particles, heat energy, electromagnetic potential energy, or the rest energy of some other particle. The reverse process of pair creation is pair annihilation, and will occur if a particle encounters its antiparticle. For example, an electron and positron in close proximity will usually destroy each other, both particles disappearing completely with the production of two, or three, photons.

When quantum and relativistic effects combine in this way, it is no surprise that relations (1.6) and (1.7) are simultaneously relevant. Combining them yields the relation

$$c/v = h/mc. \qquad (1.9)$$

This has the units of length. If m is taken to be the rest mass of a particle, then $h/m_0 c$ is a characteristic length associated with that particle, and is known as the *Compton wavelength* after Arthur Compton. For a proton it has a value of about 10^{-15} m.

Although the permanent creation of a new particle of rest mass m_0 requires an input of energy $m_0 c^2$, such a particle can be temporarily created in the absence of an energy supply. The reason for this concerns the Heisenberg uncertainty principle which allows the law of conservation of energy to be suspended for a duration Δt by an amount ΔE, where

$$\Delta E \Delta t \sim h. \qquad (1.10)$$

It follows that $m_0 c^2$ can be 'borrowed' for a time $\Delta t \sim h/m_0 c^2$ to make a temporary, or so-called virtual, particle. Of course, this temporary particle enjoys only a fleeting existence before disappearing again. It cannot, therefore, travel very far. Even at the speed of light (the upper limit) its range is $c\Delta t$, which is the quantity on the right hand side of Eq. (1.9). The Compton wavelength thus has the significance of being the maximum range of a virtual particle.

In the quantum description of the transmission of forces, messenger particles are exchanged between other particles (Fig. 5). Generally these messenger particles are virtual, and so will be limited in range to within a Compton wavelength.

In the case of electromagnetic forces, the virtual messenger particles are photons. These have zero rest mass, and hence infinite Compton wavelength and infinite range. For this reason electromagnetic forces can operate over macroscopic distances. The same is true of gravitational forces (the graviton is also massless). In contrast, the W and Z particles of the weak interaction are very massive, so the range of the weak interaction is very short ($\ll 10^{-15}$ m), and restricted to sub-nuclear distances. Similar remarks apply to the strong nuclear force between protons and neutrons.

In subatomic processes involving the emission and absorption of photons, it is necessary to consider the effects of quantum theory and electromagnetism together. This implies that the behaviour of such processes will depend on all three constants e, h and c. The speed of photons c, and Planck's constant h (and the electric constant ε), when multiplied, yield a quantity with the units of (electric charge)^2, so we may combine e, h and c to give a dimensionless ratio (a pure number) denoted α

$$\alpha \equiv e^2/4\pi\varepsilon hc = 1/137.036. \tag{1.11}$$

This quantity enters into all processes involving the interaction of matter and radiation. For example it determines the rate at which an excited atom will decay by photon emission, or the rate at which it will take up energy when immersed in a stream of photons. It also determines the degree to which atomic energy levels are split into multiplets as a result of magnetic coupling between the intrinsic magnetic moment carried by the electron and its orbital magnetic field. This so-called fine structure is apparent in the spectral lines from excited atoms. For this reason the ratio (1.11) is called the *fine structure constant*, but its significance is far more general than the name suggests.

Weak and strong nuclear processes are so short-ranged that they always operate at the quantum level. As for electromagnetism, so may the weak and strong coupling constants g_w and g_s be combined with other constants to yield dimensionless ratios

$$g_w m_p^2 c/\hbar^3 \simeq 10^{-5} \qquad (1.12)$$

$$g_s^2/\hbar c \simeq 15, \qquad (1.13)$$

where m_p is the mass of the proton. The significance of (1.12) and (1.13) is less fundamental than that of α.

Finally, returning to the subject of gravity, the quantities $\hbar$, c and G can be combined to yield a fundamental unit of length,

$$
\left.
\begin{aligned}
l_p &= (G\hbar/c^3)^{\frac{1}{2}} \simeq 10^{-35}\ \text{m} \\
\text{or of time,} & \\
t_p &= (G\hbar/c^5)^{\frac{1}{2}} \simeq 10^{-43}\ \text{s,}
\end{aligned}
\right\} \qquad (1.14)
$$

known respectively as the Planck length and Planck time. On general grounds it might be expected that at these length and time scales the effects of quantum gravity are manifested. As l_p and t_p are nearly twenty orders of magnitude beyond current experimental access, this expectation cannot be directly verified.

It is interesting that t_p supplies yet another fundamental unit of time to place alongside the age of the universe t_H and the characteristic nuclear time t_N. Their ratios are about

$$
\left.
\begin{aligned}
t_N/t_p &\sim 10^{20} \\
t_H/t_N &\sim 10^{40}.
\end{aligned}
\right\} \qquad (1.15)
$$

1.4 Subnuclear structure: a survey of fundamental particles

In spite of their basic importance for the structure of atomic nuclei, protons and neutrons are not the most elementary building blocks of nuclear matter. Evidence from the bombardment of nuclear particles by high energy electrons, and other projectiles, indicates that protons and neutrons are, in fact, composite bodies. Although the internal structure of these particles is only partially understood as yet, a coherent picture seems to be emerging.

Each proton is evidently the union of three smaller particles, called quarks. The proton contains two distinct types – or flavours – of quarks: two so-called 'up' or u quarks, each with an electric charge of $\frac{2}{3}e$, and a 'down' or d quark, with a charge $-\frac{1}{3}e$. The terms up and down are simply labels, and have nothing to do with vertical orientation. (Similar remarks apply to the other quark labels below.) The neutron is the union of one u and two d quarks. When a neutron decays into a proton, one of the d quarks changes into a u quark, an electron being created to carry away one unit of negative charge. Thus the weak nuclear force is capable of changing the flavour of quarks.

The masses of the quarks are not really known but are likely to be considerably larger than one-third of a proton mass. This is because the quarks are very strongly bound, and so relinquish a large fraction of their mass as binding energy (see page 15).

The nature of the interquark force is still not well understood. The force that 'glues' the quarks together is very strong, and it is now apparent that the strong nuclear force, which in turn binds the neutrons and protons together in nuclei, is really only a vestige of this much stronger internal 'glue'. In accordance with the concepts of quantum field theory, the interquark force is envisaged as being due to the exchange of still other types of field quanta, or particles, which are usually referred to as gluons. In the favoured theory of the glue force, called quantum chromodynamics, there are eight different flavours of gluon. Although the gluons are massless, the interquark force is still short-ranged. The reason for this is that the gluons attract each other with the same very strong force with which they attract the quarks. This is in contrast to the photon, which is not electrically charged, but merely mediates the electromagnetic force between charged particles.

Most particle physicists are of the opinion that the interquark force actually increases with the distance from each quark. If this is correct then it is impossible for an assemblage of quarks to be pulled apart. Individual quarks can then never exist in isolation. It would not be possible, for example, to smash a proton into its three constituents. This expectation is confirmed by high energy collision experiments, which have never succeeded in breaking a proton apart.

However, it is not necessary for as many as three quarks to combine: a quark doublet is permitted. One such union of two quarks is ud̄, a u quark bound together with the anti-particle counterpart of the d quark. This system has a total charge of $+e$, but because it lacks the third quark, it is somewhat lighter than a proton. The ud̄ can be produced in the laboratory and has been known for over 30 years as the so-called pi meson, or pion, denoted π^+. The antiparticle of π^+, the doublet ūd, is denoted π^-. There is also a neutral pion, π^0, that may be considered as a union of u and ū, or d and d̄. The π^0 is very unstable, as the quark may annihilate the corresponding antiquark. After an average lifetime of about 10^{-16} s it decays into two photons:

$$\pi^0 \to 2\gamma.$$

The charged pions are more stable, because the antiquarks are not the same flavour as the quarks, so they cannot directly annihilate each other. However, the weak force can change quark flavours, so it can cause the disintegration of the charged pions. The force being so weak, the decay takes considerably longer (about 10^{-8} s).

The end product of pion decay is another meson, but not one that is composed of quarks. It is called the mu meson or muon, denoted μ. Being quarkless, it does not experience the strong nuclear force. It is subject to the weak force (which is responsible for producing it) and is also electrically charged. The appearance of μ is accompanied by a new type of neutrino, called a muon-neutrino, denoted ν_μ:

$$\pi^+ \to \mu^+ + \nu_\mu$$

$$\pi^- \to \mu^- + \bar{\nu}_\mu.$$

There is no neutral muon. To distinguish muon-neutrinos from the neutrinos associated with neutron decay, the latter are called electron-neutrinos, and denoted ν_e.

Leaving aside the mesons for the moment, two strongly interacting quarks, u and d, and two weakly interacting particles, e and ν_e (plus the four antiparticles), are sufficient to account for all ordinary matter and, along with the gluons, photons, gravitons, Ws and Zs (the messenger, or exchange particles that mediate the four fundamental forces), all the essential features of particle interactions are explained. If these particles were the only fundamental units that existed, the world would probably differ little in structure from the one that we observe.

Triplets of quarks make up the heavy particles (n and p), known as *baryons*. The e and v_e are much lighter, and known as *leptons*. Two baryons and two leptons are sufficient to build a material world closely resembling the one that exists.

Mysteriously, nature seems to have produced an over-abundance of material structures, for the two baryon/two lepton scheme is found to be replicated at least twice. At high energies, where more energy is available for creating a greater rest mass, two new, heavier, types of quark appear. Their flavours are whimsically referred to as strange and charmed, denoted s and c respectively. They combine together in triplets to give further baryons, heavier than the proton and neutron, and in doublets to give heavy mesons. For example, the strange baryons, containing at least one strange quark, are denoted Σ, Λ, Ξ and Ω. All are unstable and decay on average in less than 10^{-8} s into non-strange particles. The first dis-covered charmed particle, ψ, has a mass greater than three proton masses, and consists of a $c\bar{c}$ pair. It decays in about 10^{-20} s into pions and other particles. To pair with the strange and charmed quarks are the muon, which is like a big brother of the electron, and the muon-neutrino.

Still another level is being uncovered at higher energies. Two more quarks, top and bottom, are discerned, and a new heavy lepton, tau (denoted τ), is now known, with a mass of about 3500 times that of the electron. Presumably τ has its own neutrino, v_τ, to pair with it. The number of combinations of two and three quarks and their antiquarks taken from six flavours runs into dozens, so that the world of subnuclear matter resembles a zoo of differing particle species.

The proliferation of quarks and leptons is exacerbated when account is taken of the interquark force. The favourite theory, involving eight gluon flavours, demands no less than three different types of strong 'charge' to couple the gluons and quarks, just as electric charge couples electrons and photons. The strong charge is referred to, for lack of a better name, as colour. Thus the six known quarks come in three different colours, making 18 quark types in all. The leptons, not being strongly interacting, are colourless. Particles composed of quarks (that is all the baryons and mesons) are all subject to the strong nuclear force and are collectively known as *hadrons*.

In their continuing search for simplicity at the heart of complexity, some physicists have been dismayed at the number of known quark species, and the division of matter into hadrons and leptons. They have proposed that even the

Table 2. *The elementary particles*

	Quarks		Leptons	
	Flavour	Charge	Flavour	Charge
I	Up u	$+\frac{2}{3}$	Electron, e	-1
	Down d	$-\frac{1}{3}$	Electron-neutrino, v_e	0
II	Charmed c	$+\frac{2}{3}$	Muon, μ	-1
	Strange s	$-\frac{1}{3}$	Muon-neutrino, v_μ	0
	Top t	$+\frac{2}{3}$	Tau, τ	-1
III	Bottom b	$-\frac{1}{3}$	Tau-neutrino, v_τ	0
	?	?	?	?

All ordinary matter is built from just four elementary particles and their corresponding antiparticles, which are not shown (level I). Each quark flavour comes in three colours. For some reason, nature has duplicated this scheme, at least twice (levels II and III). The leptons interact only weakly, and usually remain isolated, but the quarks are subject to the powerful force of the gluons, and are always found bound together in composites of two or three. These combinations build an enormous variety of subnuclear particles: protons, neutrons, mesons, Σs, ψs, etc. Those built from constituents of levels II and III are very unstable, and rapidly decay to level I particles. The same is true of the leptons.

Not included in this scheme are the photons, gravitons, gluons, and vector bosons that constitute the messenger particles which transmit the forces between the quarks and leptons. Charge is measured in units of the electric charge on the proton.

quarks are composites of some new, smaller units (pre-quarks). Perhaps leptons are built out of pre-quarks too. Maybe this sequence of structure within structure is unending, so that there are no truly elementary particles at all?

A simpler picture, and one that we shall adopt here, is that quarks and leptons are the fundamental building blocks of all matter. They are structureless entities, with no internal parts, at least to the extent that is consistent with quantum gravity (see Section 2.2). It is possible that new quark flavours will be discovered in future years, though this would represent further unpleasant duplication. These details are summarized in Table 2.

1.5   A brief history of the universe

Most modern cosmology is based on the concept of a big bang origin, as outlined in Section 1.1. The present rate of expansion of the universe suggests that the creation took place sometime between 15 and 20 billion years ago, and this sort of age is confirmed by independent techniques for dating the oldest stars.

As the universe expands, so any electromagnetic radiation propagating through space is stretched, producing an increase in wavelength λ and a decrease in frequency v. The effect is manifested, for example, in a shift in the spectral lines of distant galaxies towards the red end of the spectrum – the celebrated cosmological red shift.

As remarked in Section 1.1, on a large scale (greater than the size of a cluster of galaxies, that is $\gtrsim 10^{23}$ m) the cosmic material is distributed remarkably uniformly, and the expansion rate is also uniform to a high degree. The fundamental mystery about why this should be so will form the subject of later sections. For now we simply note that the expansion of a universe which remains homogeneous and isotropic may be described by a single scale factor $a(t)$, which can be regarded as proportional to the distance between two typical clusters of galaxies. As the universe expands, $a(t)$ increases with time t.

The precise form of the function $a(t)$ depends on the large scale dynamics of the universe, which are controlled by gravity. This will be discussed in detail in Section 4.2. One may, therefore, use general relativity to compute the form of $a(t)$. The result will depend on the source term adopted. If it is assumed that the dominant gravitational effects are due to the galaxies, and t is not too large, then

$$a(t) \propto t^{\frac{2}{3}}. \tag{1.16}$$

On the other hand, if the mass-energy of the universe were dominated by radiation, then

$$a(t) \propto t^{\frac{1}{2}}. \tag{1.17}$$

In both cases the initial condition $a(t) = 0$ has been chosen to correspond to a singular origin of infinite compression at $t = 0$. These two functions both have the same general shape as the curve shown in Fig. 3.

It has already been mentioned that the universe is bathed in thermal radiation at a current temperature of about 3 K. A typical wavelength of radiation scales as $\lambda \propto a(t)$, so the temperature of the radiation falls as the universe expands:

$$T \propto a^{-1}(t). \tag{1.18}$$

The energy density of electromagnetic radiation, denoted ρ_γ, is determined by Stefan's law:

$$\rho_\gamma = \mathfrak{a} T^4,$$

so

$$\rho_\gamma \propto a^{-4}(t). \tag{1.19}$$

Here $\mathfrak{a}$ denotes the so-called radiation constant, given in terms of h, c and k in Table 3. It should not be confused with the cosmological scale factor $a(t)$.

In contrast, for the mass-energy density of matter,

$$\rho_m \propto a^{-3}(t) \tag{1.20}$$

(see Section 2.4). It follows that as $a \to 0$, $\rho_\gamma > \rho_m$ so the universe was dominated by radiation energy at early times. The symbol t_{equal} will be used to denote the epoch at which $\rho_\gamma = \rho_m$. This epoch may be determined by extrapolation using observations of the current ratio of energy densities. It is found that

$$t_{\text{equal}} \sim 10^5 \text{ years.}$$

Adopting the form (1.17) for the epoch $t < t_{\text{equal}}$ one finds that

$$T \propto 1/t^{\frac{1}{2}}, \tag{1.21}$$

so the temperature rises without limit as $t \to 0$. In addition, the expansion rate is

$$H \equiv \dot{a}/a \propto 1/t, \tag{1.22}$$

which also diverges as $t \to 0$. Thus the primeval universe was characterized by two crucial features: enormous temperatures and explosive expansion. Hence the term big bang.

The proportionality constant in (1.21) depends on the detailed structure of the cosmological material. As a rough guide,

$$T \simeq 10^{10} \text{ K}/t_{\text{sec}}^{\frac{1}{2}}, \tag{1.23}$$

where t_{sec} denotes that the epoch is to be expressed in seconds. The average energy of a typical particle due to the thermal agitation is about kT, where k is Boltzmann's constant. Clearly, as we consider earlier and earlier epochs so the physics of the cosmological material will correspond to higher and higher energies. Current particle accelerators achieve the sort of energies that would have prevailed around $t \sim 10^{-12}$ s in the primeval universe. The physics of the preceding epochs must rest largely on theoretical arguments alone.

When the temperature is high enough, the thermal energy can lead to the production of particle–antiparticle pairs. This will occur when $kT \gtrsim 2m_0c^2$. Thus, before about 1 s, electron–positron pairs were present; before 10^{-6} s proton–antiproton pairs were present, etc. It follows that, near the instant of creation, all species of particles and antiparticles were present in abundance. Then as the universe expanded and cooled, so antiparticles were annihilated with particles, and disappeared from the universe, producing a great deal of electromagnetic radiation. It is this radiation that is present today, much cooled, in the form of the cosmic background heat at 3 K.

Evidently the quantity of antimatter present in the primeval universe was not precisely equal to the quantity of matter otherwise there would be no residue of matter left over to form the galaxies. The origin of this imbalance between matter and antimatter is a subject that will be taken up in Section 4.4. The primeval universe can thus be characterized by a succession of epochs (see Fig. 6). The earliest such epoch lasted about 10^{-43} s. This was when the universe was about one Planck time in age. Inside this epoch, quantum gravitational effects would have been important, perhaps leading to severe disruption of spacetime structure (see Section 2.2). Because there is as yet no reliable theory of quantum gravity,

Fig. 6. History of the universe. The most important epochs are shown (time in seconds). The earliest moment which physical theory can sensibly describe is the Planck time $\sim 10^{-43}$ s after the initial creation event. The present structure of the universe, including the fundamental forces and the particles out of which matter is built, 'froze' out of the ultra-hot furnace that characterized the first brief flash of existence.

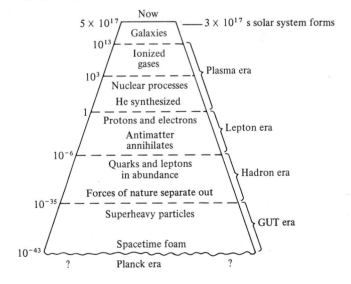

it is not possible to continue detailed investigations into this so-called Planck era.

At the end of the Planck era the temperature was presumably around 10^{32} K, and only the elementary building blocks of matter were present. The density was a colossal 10^{97} kg m^{-3}. As the temperature fell, so the building blocks (quarks?) created hadrons. One by one, as the temperature dropped further, so the vast majority of these hadrons were annihilated with their antiparticles. Of those that avoided annihilation most, being unstable, eventually decayed anyway.

After about a microsecond, the primeval cosmological material consisted of only the lighter particles: protons, neutrons, electrons, muons, pions and their antiparticles, as well as neutrinos, photons and gravitons. Heavy particles, such as those containing strange and charmed quarks, had disappeared by then. After about 10^{-6} s the temperature was too low to sustain the antiprotons and antineutrons. At around 10^{-3} s the muons had been annihilated. Finally, around 1 s, the positrons had been annihilated, leaving only neutrons, protons, electrons, neutrinos, photons and gravitons.

As the temperature continued to fall, so the thermal energy dropped below the binding energy of compound nuclei, enabling neutrons and protons to fuse together. Calculations indicate that this primeval nucleosynthesis produced about 25 per cent by weight of helium, the remaining matter being almost entirely free protons (hydrogen). After a few minutes, the temperature was too low for further fusion reactions, and very little of the nuclear material managed to assemble into nuclei heavier than helium in the short time available.

Further progressive cooling took place, but the rate of cooling diminished with time ($dT/dt \propto t^{-\frac{3}{2}}$), so that it took about 10^5 years before the temperature fell to around 10^4 K, at which point kT dropped below the ionization energy of hydrogen, and the free protons and electrons combined to form atomic hydrogen. At this stage the cosmological material became transparent to light, so that thereafter the matter and radiation were largely decoupled.

Eventually the cooling gases clumped together to form protogalaxies. Regions of enhanced density attracted further matter, thereby increasing their gravitating power. Slow contraction of these protogalaxies occurred under their own gravity. Successive stages of fragmentation followed, until the individual blobs of gas were the size of distended stars. Because these blobs were contracting rather than expanding, the general cooling tendency of the universe was more than compensated for. The contraction of the gas caused a progressive heating of the blobs, until the central temperatures rose enough to initiate nuclear reactions (several million degrees). With the onset of a nuclear energy supply, the contraction ceased as the central temperatures and pressures rose to balance the squeeze of gravity. Finally the blobs settled down to form the objects we now call stars.

A central feature of the big bang scenario given here is the assumption of thermodynamic equilibrium. Without this assumption, the description of the detailed processes in the primeval phase would have been much more complicated. Thermodynamic equilibrium implies that the cosmological material can be characterized by a single parameter – the temperature T. But is the assumption reasonable?

Measurements of the cosmic background heat radiation confirm that, to a good approximation, this primeval relic has a Planck spectrum, indicative of thermal equilibrium (see Fig. 7). However, this radiation only carries the imprint of the cosmological condition at about 10^5 years, when matter and radiation decoupled. What can be said about earlier epochs?

Because the early universe was very hot and dense, the various subatomic particles would have interacted strongly with each other, encouraging the establishment of equilibrium. On the other hand, the early epochs were also the ones of most rapid expansion, which had a tendency to disrupt equilibrium. As a general guide, if a typical reaction rate for some interactive process between two particles was much faster than the expansion rate at a particular epoch, then equilibrium will have prevailed. Reaction rates are proportional to the number density of each species of particle. For particles of equal abundance, the rate is therefore proportional to a^{-6}, which is proportional to t^{-3}. On the other hand, the expansion rate $\dot{a}/a \propto t^{-1}$. Hence, unless the other factors affecting the reaction rate are very strongly temperature – hence time – dependent (which may possibly have been the case before 10^{-35} s), it follows that, as $t \to 0$, the reaction rates exceed the expansion rate.

We may conclude, on these general grounds, that the cosmological material started out in a condition of thermodynamic equilibrium. Then, as the expansion proceeded, the reaction rates fell, and could no longer compete with the pace of expansion. Thus, the various component species of material one by one came out of equilibrium with each other. For most species it was at this stage that prolific particle – antiparticle annihilation occurred.

Consider as an example electrons and positrons. Before

Fig. 7. The energy spectrum for black body radiation has a characteristic shape. The curve is the theoretically computed graph corresponding to a temperature of 2.7 K, and the points represent the results of several observations of the cosmic microwave background radiation.

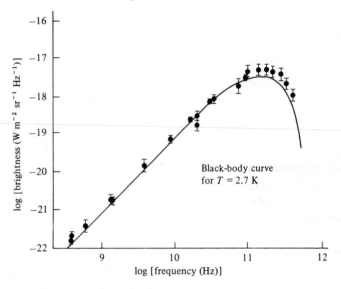

about one second these particles were in equilibrium with electromagnetic radiation (photons). Annihilations occurred at a certain (rapid) rate, but new e^+, e^- pairs were created from the heat radiation at the same rate to maintain equilibrium concentrations of each species:

$$e^+ + e^- \underset{\text{balance}}{\rightleftarrows} 2\gamma.$$

After the temperature dropped below about 10^{10} K, the photons had too little energy on average to create e^+, e^- pairs. Hence the leftward directed arrow in the above process was removed, and the rapidly depleting e^+, e^- stock could not be replenished. After a brief interval, only a tiny residue of excess electrons remained.

In thermodynamic equilibrium, the relative abundances of two species of particles is determined by Boltzmann's statistical theorem. If two states have energies E_1 and E_2, this theorem predicts that they will be populated in relative abundance in the ratio $\exp(-E_1/kT) : \exp(-E_2/kT)$. If particles of one species have rest mass m_1, while those of another species have rest mass m_2, then using $E = mc^2$ and the fact that, in equilibrium, these particles are created at the same rate as they are annihilated, the relative abundances of the two species is $\exp[(m_2 - m_1)c^2/kT]$. This 'Boltzmann factor' therefore favours particles of lower mass:

$$\frac{\text{abundance of species 1}}{\text{abundance of species 2}} = \exp[(m_2 - m_1)c^2/kT] > 1$$

if $m_2 > m_1$.

A good example concerns protons and neutrons. The neutron is slightly heavier, so would have been less abundant in the hot primeval universe. That is why the universe today is made predominantly (about 90 per cent) of protons, mainly in the form of hydrogen. The neutron – proton ratio reflects the conditions during the first second of the big bang.

The general theory of relativity also makes predictions about the future of the universe. Inspection of Fig. 3 shows that the expansion of the universe is gradually slowing in rate. This deceleration is easy to understand. The gravity of all the galaxies and other cosmic material acts as a restraint on the outward motion. If this restraint is strong enough, it will eventually (for large t) reduce the expansion to zero, after which contraction will set in, and the universe will fall back in on itself at an accelerating rate. (We ignore here the effect of the cosmic repulsion discussed in Section 1.2.) After many billions of years the galaxies would be smashed together, and any given volume of space would shrink catastrophically in a sort of 'big crunch' like the big bang in reverse. The outcome of this awsome implosion is unclear, but it might result not only in the obliteration of all familiar physical structures, but in the complete annihilation of the universe, including space and time. This singular termination of the cosmos would then be a symmetric reflection of its singular origin. In such a model the universe only exists for a finite duration.

The condition necessary to bring about a contraction is the presence of sufficient gravitating material in the universe, that is, it depends on there being a high enough mass-energy density for the given rate of expansion. Present observations suggest that the mass density of the galaxies is somewhat lower than the critical value for contraction to be assured, but unseen matter (for example in the form of neutrinos) could well make up the deficit. These issues will be discussed in more detail in Chapter 4.

If the gravity of the universe is unable to arrest the expansion, then the universe will presumably continue for ever. Individual stars will eventually either explode or burn out, and collapse to form white dwarfs, neutron stars or black holes. As the galaxies dim, so the black holes will swallow other material – dead stars, gas and dust, etc. – and grow larger. Exceedingly slow processes, such as the emission of gravitational radiation, will lead to the orbital decay of many systems, so that the galaxies will tend to slowly collapse into the black holes. Matter that escapes into the intergalactic spaces will eventually cool to the ambient temperature of the background radiation, which itself will cool according to the law (1.18).

Further subtle quantum effects may occur, which would cause the black holes to slowly evaporate away, leaving only a residue of radiation. Protons in the material that escapes a black hole demise might also gradually decay, ultimately into positrons (see page 104), which would begin to annihilate the remaining electrons. Whether or not complete annihilation would eventually occur depends on the details of the model. In any case, the ultimate condition of the ever-expanding universe would seem to be a tenuous and fading bath of photons, neutrinos, gravitons and perhaps a few electrons and positrons. Nothing further of interest would then happen for all future eternity.

2

Scales of structure

In the previous chapter the hierarchy of natural structures was outlined. Curiously the very smallest length and time scales (l_P and t_P), and the very largest (t_H) are determined by gravity. In between are the structures dominated by nuclear forces ($\lesssim 10^{-14}$ m), and electromagnetism, which controls all the structures from atomic to domestic length scales.

Although nature supplies an enormous variety of physical structures, some of them are more or less uniformly found throughout the universe with rather narrowly-defined properties. Atoms and stars are good examples. The specific details of the individual systems will depend on the laws of motion and the various boundary or initial conditions which together completely specify their behaviour. However, the gross features – size, mass, lifetime, etc. – are often determined to within an order of magnitude or so purely by the values of the fundamental constants such as G, h, c, e and m_p. Thus, for example, atoms, the structure of which depends on electromagnetism and quantum physics, are fashioned largely by the parameters e and h associated with these branches of physics. Stars, on the other hand, are gravitating objects which confine their energy electromagnetically: their structure also depends on G.

In the coming sections the principal common natural systems will be analyzed in this rather crude fashion to discover the parameters which are most crucial to their structure. It turns out that remarkably few such parameters are necessary for a fairly complete description of nature.

2.1   The role of constants in physical theory

The technique of guessing the essential features of a system without going through the detailed theory of the physical laws is often called 'dimensional analysis'. A simple example is the following. What is the period of a pendulum? To tackle this problem one must first guess the quantities that are relevant to the system under consideration. The oscillations of a pendulum on Earth are driven by the Earth's gravity, the strength of which can be characterized by the quantity g, the rate at which freely falling objects accelerate towards the ground at the Earth's surface ($g = 9.81$ m s^{-2}). Two quantities parameterize the pendulum itself: the length and the mass. Since we are interested in the period of the pendulum, we wish to construct a quantity with the units of time. Denoting time by T, length by L and mass by M, g has the units LT^{-2}. To combine g with the length L of the pendulum, and the mass M to obtain a quantity with the units T we must take the combination $(L/g)^{\frac{1}{2}}$; M is not needed. The result – that the period of a pendulum is $(L/g)^{\frac{1}{2}}$ – is not likely to be exactly correct numerically, but should still indicate the period to within an order of magnitude or so. This is confirmed by detailed theory, which yields $2\pi(L/g)^{\frac{1}{2}}$, so the answer obtained purely from dimensional arguments is only in error by a factor of about 6. For our purposes, errors of this sort are not relevant.

It is important to distinguish between quantities that are constant purely in the sense that they do not change, and fundamental universal constants. The mass of the Earth, for example, is approximately constant, but other planets exist with masses very different from the Earth. On the other hand the mass of an electron is the same for *all* electrons, wherever they are located in the universe. In the example of the pendulum, neither of the two relevant quantities L and g is fundamental, which merely reflects the fact that pendula can have periods of arbitrary duration. A pendulum is not one of the conspicuous structures that occur throughout the universe.

The number of known truly fundamental universal constants is actually rather small. We have already encountered six: c, e, h, G, g_w, g_s.

Other universal subatomic constants concern the masses of various particles: that of the electron has already been mentioned. The problem here is that, at the present level of our understanding, it is not known which particles are truly fundamental. Isolated neutrons, for example, decay to protons, electrons and antineutrinos. One should not regard the neutron, nor the many other unstable, subnuclear particles that are now known, as fundamental. Probably the electron should be accompanied by the quarks for this distinction, but the masses of the individual quarks (believed to come in at least six varieties) are quite uncertain. Moreover, when the quarks combine, their attraction is so ferocious that they shed an appreciable fraction of their mass when they glue together. Hence, it is not possible to infer from observation of, say, a proton, just what the masses of the individual quarks are. A further complication is that most quark theories treat the quark union as completely unbreakable; isolated quarks are regarded as impossible.

If quarks are permanently confined inside particles such as protons, it makes more sense to regard the proton mass, rather than the quark masses, as a fundamental unit. Hence we adopt m_e and m_p, the masses of the electron and proton respectively, as two basic natural mass units, because these are the two stable particles from which all ordinary matter is built. If recent theories suggesting that the proton is weakly unstable are correct, there is still good justification for retaining the proton mass as a fundamental unit. The proton is still by far the most stable heavy particle.

In Table 3 are listed all the universal constants that seem to be needed to account for the gross features of most known physical structures. For completeness we also list Boltzmann's constant k, which is a conversion factor between units of heat energy and temperature.

If the contents of this table are to have the universal significance claimed for them, then it is vital to verify that they are truly *constant*. If the charge on the proton, for example, were to differ from place to place, or epoch to epoch, it would not be regarded as a fundamental quantity. Some new law would be necessary to describe these variations, and this law would in turn involve its own, more fundamental, parameters.

A variety of experiments has been performed to test the

Table 3. *List of fundamental constants and derived quantities*

Name	Symbol	Numerical Value (SI units)
Charge on proton	e	1.60×10^{-19}
Planck's constant	h	6.63×10^{-34}
Speed of light	c	3.00×10^{8}
Newton's gravitational constant	G	6.67×10^{-11}
Rest mass of proton	m_p	1.67×10^{-27}
Rest mass of electron	m_e	9.11×10^{-31}
Weak force constant	g_w	1.43×10^{-62}
Strong force constant	g_s	15
Hubble constant	H	2×10^{-18}
Cosmological constant	Λ	$< 10^{-53}$
Cosmic photon/proton ratio	S	10^{9}
Permittivity of free space	ε	8.85×10^{-12}
Boltzmann's constant	k	1.38×10^{-23}
******************		
Planck length, $(Gh/c^3)^{\frac{1}{2}}$	l_P	1.62×10^{-35}
Planck time, $(Gh/c^5)^{\frac{1}{2}}$	t_P	5.39×10^{-44}
Planck mass, $(hc/G)^{\frac{1}{2}}$	m_P	2.18×10^{-8}
Proton Compton wavelength, $h/m_p c$	l_p	1.32×10^{-15}
Proton (nuclear) Compton time, $h/m_p c^2$	t_N	4.41×10^{-24}
Hubble time, H^{-1}	t_H	5.00×10^{17}
Hubble radius, cH^{-1}	r_H	1.5×10^{26}
Bohr radius, $4\pi\varepsilon h^2/m_e e^2$	a_0	5.29×10^{-11}
Radiation constant, $\pi^2 k^4/15c^3 h^3$	a	7.56×10^{-16}
Electromagnetic fine structure constant, $e^2/4\pi\varepsilon hc$	α	7.30×10^{-3}
Weak fine structure constant, $g_w m_e^2 c/h^3$	α_w	3.05×10^{-12}
Gravitational fine structure constant, Gm_p^2/hc	α_G	5.90×10^{-39}

The fundamental constants of nature listed here largely determine the essential features of most known physical structures. Many of these features are remarkably sensitive to the values of the constants, and to certain apparently accidental numerical relations between them. Note that H (and probably Λ) are not actually constant, but vary over cosmological time scales, while k and ε are merely conversion factors between two systems of units.

'constancy of the constants'. These can be divided into two classes: local experiments and cosmological observations. The local experiments look for relics left by the effects of variation over geological time scales. For example, changes in g_s or e would show up in nuclear stability and radioactive lifetimes for α-decay. Variations of G affect the luminosity of the sun and the Earth's orbital motion, and could be expected to leave traces in the geological record.

The cosmological observations can probe both spatial and temporal variations in physics, because the remote regions of the universe are viewed today in the light they emitted billions of years ago. Variations in e or m_e would affect the details of spectra of distant galaxies. Changes in G should produce noticeable evolutionary effects on galactic structure, etc.

None of these careful analyses provides any compelling evidence in favour of variations in the fundamental constants. Some writers maintain that there is some evidence for a variation of G by less than one part in 10^{10} per year, but the data concerned are open to alternative explanations.

Three exceptions concern the cosmological parameters H, Λ and S. As already remarked, H is not intended to be constant: H^{-1} is roughly the age of the universe. The cosmological term Λ is believed to vary exceedingly weakly with time at this epoch. However, during the very early stages of the universe, variations in Λ might well have been dramatic and significant. This topic will be discussed further in Section 4.5.

The ratio S is clearly not precisely constant because photons are continually being created and absorbed. There is a steady accumulation of heat and light from stars, for example. Nevertheless, the primeval photon content of the universe is far larger than the starlight content, so that variations in S, even over cosmological times, are small.

2.2   Microstructures

In the following section it will be shown how the scales of the principal microscopic structures in the universe are indeed determined by the constants listed in Table 3. In this section and the two that follow the presentation is based on the review of Carr & Rees (see bibliography).

Spacetime foam

The smallest structures predicted by known physics occur at the Planck length, $l_p \sim 10^{-35}$ m. Although the physics at such ultra-small dimensions is hopelessly beyond current experimental access, some theoretical modelling suggests that at scales $\sim l_p$ profound modifications are necessary to the traditional concepts of space and time. The Planck regime is one that is characterized by the relevance of both gravity and quantum physics. As remarked in Section 1.3, Einstein's general theory of relativity describes gravity as a distortion or curvature of spacetime. One of the basic features of the quantum theory is the way in which particles and fields can undergo spontaneous fluctuations of a random nature. Hence, in the regime of quantum gravity, expected to become important on Planck length and time scales, it seems likely that violent fluctuations occur in the curvature of spacetime. Indeed, it is even possible that the topology of spacetime is very complex, with 'wormholes' and 'bridges'. It has been remarked that spacetime, which is often compared to a smooth sheet or canvas on which nature's activity is painted, resembles more a sponge-like, or foamy structure, on these very small length and time scales. Another picture is that spacetime is, in some vague sense, composed of Planck-sized black holes packed together.

None of these images amounts to more than a rough and schematic expression of the rather abstract concepts that characterize the quantum theory of gravity, itself an incomplete and unsatisfactory theory.

Nuclei

The nuclei of atoms involve protons and neutrons bound together by the strong nuclear force. One might therefore expect the masses of these particles, and features of the nuclear force, to determine the size of nuclei. The proton and neutron masses differ by less than 0.1 per cent so the situation can be described in terms of the mass m_p only. Since the nucleus is a quantum system, h will play a role. Also, because the nuclear force is strong, a significant fraction of the mass (about 1 per cent) is sacrificed as binding energy, so relativity is also important.

The constants, m_p, h and c can be combined into a quantity with the unit of length (the Compton wavelength of the proton – see Section 1.3):

$$h/m_p c \sim 10^{-15} \text{ m.}$$

It is often referred to as the size of a proton, though that designation is a little misleading. It is more accurate to say that it would not be possible to localize a proton in a region of space smaller than $h/m_p c$. The reason for this concerns the Heisenberg uncertainty principle in the form

$$\Delta p \Delta x \sim h. \tag{2.1}$$

This relation implies that if the position x of a particle is determined to within a range of values Δx, then its momentum p is uncertain by an amount $\Delta p \sim h/\Delta x$. Thus, if $\Delta x \sim 10^{-15}$ m, then $\Delta p \sim 10^{-18}$ kg m s^{-1}. For a proton, such a momentum is only achieved by travelling close to the speed of light. The kinetic energy then becomes so great that it is sufficient to create new particles:

$$\text{energy} \rightarrow \text{p} + \bar{\text{p}},$$

where a new proton–antiproton pair appears. Evidently the original proton, if squashed into a region $\sim 10^{-15}$ m, would start to breed other identical protons (and antiprotons), so that its original identity would be lost. Clearly the size of atomic nuclei must exceed the proton's Compton wavelength. In fact, small atomic nuclei are not much larger than this.

Nothing has yet been said about the effect of the nuclear force that acts between the protons and neutrons. It turns out that the strength of the nuclear force, as determined by g_s, is not a relevant parameter as far as nuclear size is concerned. This is because nuclear forces only act between nearest neighbours, and therefore display the property of saturation: a nucleus with many particles is not more tightly bound than one with a few particles. Each particle simply sticks to its nearby neighbours. The size of the nucleus is therefore essentially determined by the total number of nuclear particles and the range of the force between individual particles. As remarked in Section 1.2, the force falls rapidly to zero beyond about 10^{-15} m, so the range, and hence the average separation between the protons and neutrons, is actually about the same as $h/m_{\mathrm{p}}c$.

The reason for this coincidence can be explained with a rather simple model for the origin of the nuclear forces originally proposed in the 1930s by Hideki Yukawa. In this model the force between nuclear particles is described in the language of quantized fields, as explained in Section 1.3. The force arises from the exchange of 'messenger' particles between protons and neutrons (see Fig.5). Since the force is short-ranged, the messenger particles must be massive, because the range is roughly the Compton wavelength of the messenger particles. The above coincidence then suggests messenger particles with a rest mass not very different from that of protons. This turns out to be correct, and one may identify the messengers with the pi mesons, which have masses $\sim 0.1m_{\mathrm{p}}$. The coincidence is therefore simply a result of the fact that the nuclear force messengers, and the nuclear particles, are both constructed from the same quarks.

Atoms and molecules

Next up on the scale of structures come atoms. The electron cloud is bound to the nucleus by electromagnetic forces, so one expects atomic size to be determined by e, h and m_{e}. These can be combined into a quantity with the units of length (the so-called Bohr radius) as follows:

$$a_{\mathrm{o}} = 4\pi\varepsilon h^2/m_{\mathrm{e}}e^2 \sim 10^{-10}\ \mathrm{m} \tag{2.2}$$

and into a unit of energy

$$E_{\mathrm{atom}} = m_{\mathrm{e}}e^4/16\pi^2\varepsilon^2 h^2 \sim 10^{-18}\ \mathrm{J}. \tag{2.3}$$

Exact calculations yield, for example, an energy of $-m_{\mathrm{e}}e^4/32\pi^2\varepsilon^2 h^2$ for the lowest energy state of the hydrogen atom. (The minus sign is an expression of the fact that the electron is bound in the hydrogen atom, so that energy has to be expended to remove it.) Experiment verifies that atoms have diameters in the region of 10^{-10} m.

Molecules form from the residual electromagnetic effects which bind the electrons to the nuclei. These interatomic forces are fairly short in range, so that molecular structures consisting of a small number of atoms are comparable in size to the atoms themselves, that is, interatomic separations are not much greater than atomic size. Interatomic bonds due to these residual effects are somewhat weaker than atomic bonds. Roughly

$$E_{mol} \sim 0.1 E_{atom}. \tag{2.4}$$

2.3 Macrostructures
Solid bodies

When considering matter in bulk the crucial parameter is the temperature, because this determines whether the material is solid, liquid or gaseous. At a temperature T, the average thermal energy of a molecule is $\sim kT$, so the requirement for solidity is $kT \lesssim 0.1 E_{atom}$ if it is assumed that the molecular binding energy is due to a roughly 10 per cent residue of the atomic binding energy. Using (2.3) one finds $T \lesssim 10^4$ K, so there is plenty of opportunity in our universe, the background temperature of which is a mere 3 K, for solid objects.

When atoms are packed together in bulk, the interatomic forces are comparable to the binding energies of the peripheral electrons, and in many cases, such as metallic crystals, it is more accurate to envisage these peripheral electrons detaching themselves completely from the atoms and wandering about freely in the bulk of the material. In any case, the rigidity properties of familiar materials can be largely attributed to the electrons.

Macroscopic matter will exist at a certain density because of a balance of forces. The forces of electromagnetism and gravity will try to hold the material together, but this implosive tendency will be counterbalanced by an outward pressure due, in the case of solids, to the Pauli exclusion principle. This

Table 4. *Scales of structure*

System	Size (m)	Structural feature	Mass (kg)	Characteristic time (s)
Quantized gravity	10^{-35}	Spacetime foam	10^{-8}	10^{-43}
Quarks, leptons	$<10^{-18}$	Elementary, structureless particles	?	$<10^{-26}$
Nuclear particles	10^{-15}	Union of quarks	10^{-27}	10^{-24}
Nucleus	10^{-14}	Union of nuclear particles	10^{-25}	10^{-23}
Atom	10^{-10}	Nucleus and electrons	10^{-25}	10^{-16}
Biological molecule	10^{-7}	Union of atoms	10^{-20}	10^{3}
Living cell	10^{-5}	Complex order	10^{-10}	10^{3}
Advanced life form	1	Intelligent organization	10^{2}	10^{9}
City	10^{4}	Social order	10^{11}	10^{9}
Mountain, asteroid	$10^{4}-10^{5}$	Irregular	$10^{12}-10^{13}$	—
Planet	10^{7}	Gravitationally dominated	10^{24}	10^{4}
Star	10^{9}	Nuclear reactions	10^{30}	10^{17}
Planetary system	10^{11}	Star and planets	10^{30}	10^{8}
Star cluster	10^{18}	Gravitationally bound	10^{35}	10^{15}
Galaxy	10^{21}	Nucleus and spiral arms	10^{41}	10^{16}
Cluster of galaxies	10^{23}	Largest known structure	10^{43}	10^{17}
Universe	10^{26}	Uniformity	10^{53}	10^{18}

The table shows the principal steps in the structural hierarchy that builds our physical universe. Numbers quoted are taken to the nearest power of 10. The characteristic time is chosen to be the shortest duration over which the system transmits appreciable information or would undergo major structural change. For the first four entries this is the light travel time across the system. In the case of biological and social systems it is the reproduction or growth time. For stars, the average lifetime is given, but for the other gravitationally bound systems the free fall time (roughly the time for the system to implode under its own gravity) is more appropriate. The entry for the atom is the electron orbital time. The final entry refers to the age of the universe.

principle states that no two electrons may occupy the same state, which crudely speaking implies that there is a sort of repulsion that tries to keep electrons apart from each other, quite independently of the electric repulsion.

To illustrate the Pauli principle at work, we shall consider an elementary one-dimensional problem. Suppose that a single electron is confined in a rigid box of length L. According to quantum theory, the electron's behaviour is described by a wave, whose wavelength λ is related to the electron's momentum p by Eq. (1.8). Because the electron is strictly confined to the box, the wave amplitude is zero outside the box, so it must vanish at the ends of the box to remain continuous: it cannot penetrate into the region beyond. The longest wavelength, hence the minimum momentum, will thus be achieved when exactly one-half a wave fits into the length L, that is, when $\lambda/2 = L$ (see Fig. 8). An electron in this 'ground' state has the minimal attainable kinetic energy $E = p^2/2m_e = h^2/8m_e L^2$.

Because of the Pauli principle, if a second electron is placed in the box (we ignore their electric repulsion), it cannot reside in this ground state. Instead, the lowest attainable energy will be the first excited state, achieved when one whole wave fits into the length L. In this case $\lambda = L$ and $E = 4 \times (h^2/8m_e L^2)$.

Fig. 8. Waves in a box. This one-dimensional example shows how a wave confined to a box of length L can only adopt a discrete set of wavelengths $\lambda = 2L$, $\lambda = L$, $\lambda = 3L/2$, etc. Because of the Pauli principle, electron waves in such a box must adopt each successive wave configuration in turn. No two electrons can have the same wave form. (Electron spin is ignored.)

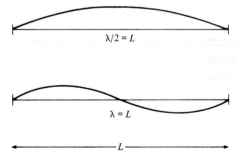

$\lambda/2 = L$

$\lambda = L$

L

Continuing in this way, successive electrons must occupy higher and higher energy levels, involving waves of shorter and shorter length. The kth electron will have kinetic energy $k^2 h^2 / 8 m_e L^2$.

If the total number of electrons is N, then the total energy can be computed by summing $k^2 h^2 / 8 m_e L^2$ for k running from 1 to N. The result is, for large N, approximately $N^3 h^2 / 24 m_e L^2$, so the average energy per electron is $N^2 h^2 / 24 m_e L^2$. Expressed in terms of the number density of electrons n (that is the number per unit length in this one-dimensional case) the Pauli exclusion principle requires an average energy per electron of about $n^2 \hbar^2 / m_e$. This repulsive effect is usually referred to as electron *degeneracy pressure*.

In passing to three dimensions, the result is the same, to within small numerical factors, except that if n is now a number density of electrons per unit volume, we must replace n^2 by $n^{\frac{2}{3}}$. Therefore

$$E_{\text{degeneracy}} \sim n^{\frac{2}{3}} \hbar^2 / m_e. \qquad (2.5)$$

In a small body, where gravity is negligible, this degeneracy pressure will be balanced by the electric forces of attraction between the electrons and the positively charged nuclei in the material. The two effects will come into equilibrium when the energy of attraction per electron is comparable to $E_{\text{degeneracy}}$. In a volume of electrically neutral material, each electron will be accompanied by an equal and opposite number of positive charges from the nuclei. If the average separation between electrons and nuclei is of order a_0 (given by Eq. (2.2)), then the magnitude of the electrostatic energy per electron will be about $e^2 / 4 \pi \varepsilon a_0$, because each neighbouring attracting pair of charges will approximately neutralize each other's forces on the remaining particles. If the average separation is a_0 then the total number of electrons per unit volume, n, is $1/a_0{}^3$, so we may write

$$E_{\text{electric}} \sim e^2 n^{\frac{1}{3}} / 4 \pi \varepsilon. \qquad (2.6)$$

Equating (2.5) and (2.6) yields the number density $n = e^6 m_e{}^3 / 64 \pi^3 \varepsilon^3 \hbar^6$. To convert n into a mass density, we note that most of the mass of a solid is due to the nuclei, the number density of which is also n. Taking the nuclear mass as a few times m_p, and ignoring factors of order unity, we have

$$\text{mass density} \sim e^6 m_e{}^3 m_p / 64\pi^3 \varepsilon^3 \hbar^6 \sim m_p / a_0{}^3 \sim 10^3 \text{ kg m}^{-3}$$

which is a very realistic value.

Planets

The above calculation must be modified if the body concerned is massive enough for gravity to become important. Although the gravitational forces between individual atoms are nearly 40 powers of 10 weaker than electric or degeneracy effects, unlike electricity, gravity is not neutralized by neighbouring particles so it is cumulative with increasing number of particles N. For a spherical body of mass M and radius R, the gravitational energy is $\sim -GM^2/R \sim -GA^2 m_p{}^2 N^2/R$, where A is the molecular weight of the material. For large solid bodies, such as the Earth, A may be quite large, so we include it explicitly as an additional factor here. This becomes comparable with electric and degeneracy effects when

$$GA^2 m_p{}^2 N^2 / R \sim Ne^2 / 4\pi\varepsilon a_0$$

or when

$$R \sim N a_0 A^2 (4\pi\varepsilon G m_p{}^2 / e^2). \tag{2.7}$$

The quantity $4\pi\varepsilon G m_p{}^2 / e^2$ is a dimensionless number with a fundamental significance: it is the ratio of the strength of gravitational to electromagnetic forces between protons. In analogy to the fine structure constant, (1.11), for electromagnetism, we introduce a gravitational 'fine structure' constant, defined by

$$\alpha_G \equiv G m_p{}^2 / \hbar c \simeq 5 \times 10^{-39}. \tag{2.8}$$

This fundamental dimensionless number will recur frequently in the coming sections. In terms of α and α_G, (2.7) becomes

$$R_{\text{planet}} \sim (\alpha/\alpha_G)^{\frac{1}{2}} a_0 / A, \tag{2.9}$$

where use has been made of $N \sim R^3 / a_0{}^3$.

For a fairly dense object, $A \simeq 50$, so

$$R_{\text{planet}} \sim 10^7 \text{ m}$$

which is comparable to the radius of the Earth. We may conclude that in such a body, gravitational forces will cause considerable modification to the solid structure of the material, resulting in appreciable compression and liquefaction. This is, of course, the case with the Earth.

The acceleration due to gravity at the body's surface is

$$g = GM/R^2. \tag{2.10}$$

If a 'bump' (mountain) of height H forms on the surface, this will exert a pressure at its base, which if strong enough will melt the underlying material, enabling the mountain to sink. The available potential energy is $\sim HAm_p g$ per molecule, and if this becomes an appreciable fraction, say 10 per cent, of the molecular binding energy for the substance at the base, then liquefaction will occur. Such a body will automatically tend to adopt a spherical shape, like Earth. The maximum height of a mountain is therefore

$$
\begin{aligned}
H_{\max} &\sim (10^{-2} e^4 m_e / 16\pi^2 \varepsilon^2 \hbar^2)/(GMm_p A/R^2) \\
&\sim 10^{-2} (\alpha/\alpha_G)(a_0{}^2/A^2 R), \tag{2.11}
\end{aligned}
$$

where we have used (2.4), and the fact that $M \sim Am_p R^3/a_0{}^3$. Substituting $R = R_{\text{planet}}$ from (2.9)

$$
\begin{aligned}
H_{\max} &\sim 10^{-2} R_{\text{planet}} \\
&\sim 10^5 \text{ m}, \tag{2.12}
\end{aligned}
$$

which is a very reasonable estimate. Earth's largest mountain is nearer $10^{-3} R_{\text{planet}}$, which is clearly considerably less than $H_{\max}$. Mars' largest mountain is rather closer to $H_{\max}$.

An irregular-shaped body can form if its mass is somewhat smaller than that of a planet. Irregularities will be large when the height H of a 'mountain' becomes comparable to the mean radius R of the body. Such a structure will only be supported if $HAm_p g \lesssim 0.1 E_{\text{mol}}$. Using (2.4) and (2.10) and putting $H \sim R$ yields

$$R \lesssim 0.1 R_{\text{planet}} \sim 10^6 \text{ m}.$$

Irregularly shaped objects below this size are well known in the solar system: the asteroids.

Stars

The analysis given above for planets neglected the effects of heat energy. A large gravitating object compressing the central material will raise the internal temperature, and if this reaches as high as several million degrees then nuclear reactions will be initiated and the object's structure will be drastically altered. At these temperatures the body will be gaseous rather than solid, and the nuclear reactions will cause it to radiate profusely. The object will be a star.

A ball of gas of radius R will remain in equilibrium if its self-gravity is supported by the combined effort of its internal thermal pressure and its electron degeneracy pressure. This will be the case if the gravitational energy per particle is comparable to the thermal energy kT plus the degeneracy energy, (2.5). For hydrogen gas this implies

$$kT + N_*^{\frac{2}{3}} \hbar^2 / m_e R^2 \sim GMm_p/R \sim Gm_p^{\,2}N_*/R, \quad (2.13)$$

where N_* is the total number of protons in the star.

At low density (large R), the R^{-2} term is small – degeneracy pressure is negligible – so the temperature $T \propto R^{-1}$. This is the case when the star first forms from a slowly contracting cloud of gas. Eventually, however, as the radius shrinks, the degeneracy term becomes important, and the temperature reaches a maximum when

$$Gm_p^{\,2}N_*/R - N_*^{\frac{2}{3}} \hbar^2 / m_e R^2$$

is greatest. This occurs for $R = 2\hbar^2/Gm_p^{\,2}m_e N_*^{\frac{1}{3}}$ for which the temperature is given by

$$kT_{\text{max}} \sim (Gm_p^{\,2}/\hbar c)^2 N_*^{\frac{4}{3}} m_e c^2. \quad (2.14)$$

For the object to become an ordinary star, T_{max} must be hot enough for nuclear reactions to occur. The required temperature depends on the details of the strong and electromagnetic forces which form the nuclear potential barrier through which the thermally agitated protons must penetrate. At first it seems that this must require $kT \sim 10m_e c^2$ (the energy released during the fusion process) but in fact nuclear penetration can occur at much lower temperatures because of the quantum tunnelling effect and the fact that, at any given temperature, the individual proton energies will be distributed over a wide range, so that some will always have substantially greater energy than average. In Section 3.3 it will be shown that $kT \sim 10^{-2} m_p e^4 / 16\pi^2 \varepsilon^2 \hbar^2$.

We may use this information to compute the minimum number of protons in a star. From (2.14), putting $kT_{max} \sim 10^{-2} m_p e^4 / 16\pi^2 \varepsilon^2 \hbar^2$, one obtains

$$(10^{-2}\alpha^2 m_p/m_e)^{\frac{3}{4}}\alpha_G^{-\frac{3}{2}} \sim 0.1\alpha_G^{-\frac{3}{2}} \sim 10^{56} \qquad (2.15)$$

and a corresponding radius of

$$R_{min} \sim \alpha_G^{-\frac{1}{2}} \alpha a_0 \sim 10^{18} a_0 \sim 10^8 \text{ m}. \qquad (2.16)$$

This is comparable to the size of a planet, so we may conclude that there is no sharp division between large planets, such as Jupiter, and small stars. Indeed, the central temperature of Jupiter is probably about 25 000 K.

In the above estimates the effects of radiation were ignored. Is this reasonable? At a temperature T the energy density of radiation is aT^4, where a is the radiation constant: $a = 8\pi^5 k^4 / 15c^3 h^3$. The total radiation energy inside the star will be about $aT^4 R^3$, which should be compared to the kinetic energy of the particles $\sim N_* kT$. If we ignore the degeneracy pressure, then (2.13) gives $kTR \sim Gm_p^2 N_*$, so

$$aT^4 R^3 / N_* kT \sim \alpha_G^3 N_*^2$$

which will be of order unity when

$$N_* \sim \alpha_G^{-\frac{3}{2}}.$$

For N_* greater than this value the dynamics of the star will be dominated by radiation, and it is probable that this would lead to instabilities such as pulsation and eventual disruption. Hence we regard $\alpha_G^{-\frac{3}{2}}$ as the maximum value of

N_*. This is surprisingly close to the minimum value of N, given by (2.15), and so defines a rather narrow range of permissible values around $\alpha_G^{-\frac{3}{2}} \sim 10^{57}$. From these elementary ideas one may conclude that the number of particles contained in a typical star is given by the astonishingly simple formula

$$N_* \sim \alpha_G^{-\frac{3}{2}}. \qquad (2.17)$$

The mass is given roughly by

$$M_* \sim N_* m_p \sim \alpha_G^{-\frac{3}{2}} m_p. \qquad (2.18)$$

Whatever the conditions inside the star, what we actually see is the surface. The most important surface parameter is the luminosity, L, defined as the flow of energy per unit time into space, in the form of radiation. If the star is static, and in a steady state, then obviously the rate of loss of energy from the surface is exactly balanced by the rate of energy generation in the interior. The latter is determined by the central temperature, T_c, which controls the rate of nuclear burning in the core. There will be a temperature gradient between the core and the surface, but an order of magnitude estimate for the total radiant energy content of the star is obtained by putting $T = T_c$ in the formula aT^4R^3 used above. The luminosity is then given by the energy content multiplied by the rate at which energy migrates out through the star. In quantum language, this is the energy of all the photons in the star divided by the photon diffusion time inside the star.

The photons produced in the core do not readily pass through the stellar material, which is highly opaque. Instead they travel only a short distance before scattering from an ion or an electron. For moderately sized stars this opacity is mainly caused by scattering from free electrons. If the electrons' recoil is ignored (the photon energy is relatively low in the outer layers of the star where most of the material is situated) then the scattering cross-section is given by the Thomson formula

$$\sigma = e^4/16\pi^2\varepsilon^2 m_e^2 c^4. \qquad (2.19)$$

The mean distance travelled by a typical photon before scattering is then

$$\bar{l} \simeq 1/n_e\sigma, \tag{2.20}$$

where n_e is the average number density of electrons, which will be roughly the average density of protons, or about N/R^3. Thus

$$\bar{l} \sim 16\pi^2\varepsilon^2 m_e{}^2 c^4 R^3/Ne^4. \tag{2.21}$$

If the photon were free to escape it would take a time R/c to reach the surface. Instead, being impeded by the electrons, the photon executes a random walk inside the star, only percolating to the surface by chance. The escape time is thereby lengthened by the factor $R/\bar{l}$. The luminosity is, therefore

$$L \sim \frac{aT_c{}^4 R^3}{(R/c)(R/\bar{l})}.$$

This is usually re-expressed in terms of the opacity of the stellar material, defined as $\kappa \equiv (\bar{l} \times \text{mass density})^{-1}$. Using the fact that the mass density is $\sim M/R^3$, one finds

$$L \sim acT_c{}^4 R^4/\kappa M, \tag{2.22}$$

which is essentially (total available radiation energy)/(leakage time from star).

An alternative expression for L results from noting that, as shown above, the energy density of radiation in a typical star is not very different from the thermal energy of the particles. If degeneracy effects are neglected, Eq. (2.13) reveals that the thermal agitation energy kT is about the same as the gravitational binding energy per particle. Combining these two approximate equalities yields total radiation energy $\sim$ total gravitational binding energy or

$$aT_c{}^4 R^3 \sim GM^2/R. \tag{2.23}$$

Using (2.23) to eliminate $aT_c{}^4$ from (2.22), one arrives at the so-called Eddington luminosity

$$L \sim GMc/\kappa. \tag{2.24}$$

For our purpose it is convenient to see the explicit dependence of L on the electric charge e, so we use (2.21) to eliminate

κ from (2.22) and (2.24). The two expressions for the luminosity become

$$L \sim 16\pi^2\varepsilon^2 aT_c^4 R^4 m_e^2 c^5 / Ne^4 \qquad (2.25)$$

and

$$L \sim 16\pi^2\varepsilon^2 GMm_p m_e^2 c^5 / e^4 \qquad (2.26)$$

respectively.

Another important characteristic of a star is its lifetime. A casual inspection of the sky gives the impression of a static, unchanging universe. Except in the cases of novae or variables, the stars do not seem to change from one century to the next. In particular, the sun has remained stable, changing very little in its luminosity, for over 4 billion years. This is known because there are records of life on Earth over 3 billion years old. Liquid oceans have existed on Earth for the greater part of the history of the solar system, implying rather narrow constraints on the temperature and luminosity variability of the sun.

The fact that we live in stable environment of such great longevity is a consequence of the sun's not burning up its hydrogen fuel too rapidly. When this fuel is depleted (in about 5 billion years time) the sun will embark on an erratic career, eventually collapsing to a white dwarf, when nuclear reactions can no longer sustain it. Similar behaviour characterizes most other stars.

The rate at which nuclear fuel is consumed inside a star depends on the star's luminosity, which in turn depends both on the strength of gravity (through G) and electromagnetism (through the opacity of the stellar material, hence e). The lifetime of a star is given, crudely, by the total energy reserves divided by the rate of energy consumption, L. The former is computed from the fact that the fusion of hydrogen to heavier elements, which is the source of the stellar heat, results in the release of about 1 per cent of each proton's rest mass. If all the hydrogen in the star is consumed in this way, about $10^{-2}\, Mc^2$ will be liberated. Using (2.26), it follows that the lifetime of the star will be about

$$t_* \sim 10^{-2}e^4/16\pi^2\varepsilon^2 Gm_e{}^2 m_p c^3.$$

This relation may be expressed more conveniently in terms of the nuclear timescale, $t_N \sim h/m_p c^2$:

$$t_* \sim [10^{-2}\alpha^2(m_p/m_e)^2]t_N \alpha_G{}^{-1}. \tag{2.27}$$

The quantity in square braces is of order unity so

$$t_* \sim \alpha_G{}^{-1}t_N \sim 10^{40}t_N. \tag{2.28}$$

Once again we encounter the 'magic' number $\sim 10^{40}$. Recalling the relations (1.15) we may write

$$t_*/t_P \sim (t_N/t_P)^3. \tag{2.29}$$

Moreover we find

$$t_* \sim t_H, \tag{2.30}$$

which says that the lifetime of a typical star is comparable to the present age of the universe, which is correct. Using (2.18), one obtains a further relation

$$M_*/m_p \sim t_*/t_P. \tag{2.31}$$

It is evident from the foregoing discussion that if gravity were stronger, stars would burn out faster. An increase in G by a factor of 10 would totally alter the structure of the solar system over the time scale of its present history. The Earth, for example would no longer exist, having been vaporized as the sun approached its red giant phase at the end of its hydrogen consumption.

The extreme weakness of gravity is a remarkable feature of nature. The gravity inside a hydrogen atom is close to 10^{40} times weaker than electromagnetism, a number which we now see has a direct influence on the lifetime of the star. In short, the extremely long time scale required for major cosmic change is directly attributable to the weakness of gravity.

2.4   Cosmic structure

The most conspicuous structures above the scale of stars are the galaxies. The Milky Way, with around 10^{11}

stars of mass comparable to that of the sun, $M_\odot$, is typical. As remarked in Section 1.1, the galaxies themselves cluster together into agglomerations, with a total mass of around $10^{14} M_\odot$ and sizes of about 10^{23} m.

Unlike the structures considered in Sections 2.2 and 2.3, it does not appear possible to account for the sizes or masses of galaxies and clusters of galaxies on the basis of simple physical constants alone. When the big bang theory was first established, it was hoped that the existence of galaxies could be explained in the following way. The primeval gases emerging from the big bang were assumed to be distributed uniformly throughout space. As the universe expanded, so the local density of the gas steadily fell.

Here and there among this hot gaseous material, statistical fluctuations would result in slightly enhanced density in some regions, and slightly reduced density in others, purely as a result of random perturbations in the otherwise smooth distribution of gas. The overdense regions would then exert a slightly greater gravitational attraction on the surrounding material and thereby tend to accumulate more material from their environment. There is thus a natural tendency for the density perturbations to grow with time. Thus might the galaxies eventually form.

The growth of overdense regions has to compete with the ubiquitous cosmological expansion. The expansion tries to reduce the density of gas, while the density perturbation operates to restrain the local expansion. The result of this competition is that the denser regions of gas only grow very slowly – too slowly for them to explain the existence of galaxies in the time available since the creation.

One way to avoid this problem is to assume that substantial density irregularities were present at the outset, necessitating only a modest enhancement during the subsequent expansion. While this undoubtedly explains the existence of galaxies, it means that one must accept that the appropriate irregularities were obligingly present in the beginning, on the right scale of size, with a density great enough to produce galaxies, yet not so great as to produce total gravitational collapse to black holes. This explanation of galaxies and clusters of galaxies thus rests on the *initial conditions* rather than the fundamental constants. Had these conditions differed in size or degree of irregularity, then the large scale organization of the cosmos would have been totally different.

It should be remarked that the origin of the density perturbations has been sought in the details of the cosmological material at very early epochs (< 1 s). Many mechanisms have been suggested, involving forces other than gravity. It is conceivable that hadronic processes before 10^{-6} s could lead to clumping of matter into blobs of galactic mass. Unfortunately these early epochs are so badly understood that any attempted explanation for the origin of galaxies based on the physics at that time is extremely conjectural. We shall return to this topic in more detail in the next chapter.

The largest scale of structure is, of course, the universe itself. As described in Sections 1.1 and 1.5, the unexpectedly high degree of cosmic uniformity on the very large scale enables the global behaviour of the universe to be characterized by a single scale factor $a(t)$. From this parameter, one may build a characteristic time scale, the Hubble time $t_H \equiv a/\dot{a}$, comparable to the age of the universe.

A further dynamical parameter of interest is the rate at which the expansion of the universe is slowing down. As explained in Chapter 1, this deceleration is produced by the gravitational drag of all the galaxies and other cosmic material. Its strength therefore depends on the nature of the gravitating substances. For ordinary matter the function (1.16) applies, while in a universe dominated by radiation (1.17) is more appropriate. In Chapter 4 the cosmic dynamics will be analyzed in greater detail, where it will be shown that other considerations can also affect the form of $a(t)$.

Whatever effects one must include to determine $a(t)$, the actual deceleration can be described by the parameter

$$q = -\ddot{a}a/\dot{a}^2 \tag{2.32}$$

which takes the values $\frac{1}{2}$ and 1 for cases (1.16) and (1.17) respectively.

The gravitating power of the cosmic material, which partly determines q, can be expressed in terms of the energy density, ρ, of the various contributions. Present observations suggest that, for matter and electromagnetic radiation respectively

$$\rho_m \simeq 10^{-11} \text{ J m}^{-3} \tag{2.33}$$

$$\rho_\gamma \simeq 10^{-14} \text{ J m}^{-3} \tag{2.34}$$

and these vary with time according to (1.19) and (1.20). The energy density of the matter, (2.33), refers to the rest mass energy, which is the dominant contribution at this epoch. The quoted estimate may have to be revised upwards in the light of some recent experimental results concerning neutrinos (see Section 3.1).

The ratio ρ_γ/ρ_m varies like $a^{-1}(t)$. This is because ρ_γ suffers an additional reduction over ρ_m as the universe expands, due to the red shift. A much more significant radiation/matter ratio is provided by examining the particle density, rather than the energy density. In the case of non-relativistic matter, ρ_m is proportional to the particle density. Because most cosmic material is in the form of hydrogen:

$$\rho_m \simeq m_p n_p \propto a^{-3}, \tag{2.35}$$

where n_p is the number density of protons.

On the other hand,

$$\rho_\gamma \simeq n_\gamma h\nu \tag{2.36}$$

where n_γ is the density of photons and $h\nu$ is the energy of a typical photon (see Eq. (1.7)). As the universe expands, so the wavelength of radiation scales like $a(t)$, and hence $\nu \propto a^{-1}$. Noting that $\rho_\gamma \propto a^{-4}$, one obtains from (2.36)

$$n_\gamma \propto a^{-3}. \tag{2.37}$$

It follows that n_γ/n_p is a constant, independent of time. It is therefore a fundamental dimensionless ratio of great importance to cosmology, and it has been denoted by S. From (2.33) and (2.34) one deduces

$$S \equiv n_\gamma/n_p \sim 10^9. \tag{2.38}$$

Although S is specifically the photon/proton ratio, there are undoubtedly other contributions (gravitons, neutrinos) to the radiation content of the universe. Most cosmologists believe that they are comparable to the photon content (see Section 4.4).

Another way of parameterizing Eq. (2.33) is to adopt a natural unit of volume, and construct a particle number, rather than a density. An obvious cosmic distance scale is provided by ct_H, the distance travelled by light in one Hubble time. As this is roughly the distance travelled by light since the creation of the universe, it represents a sort of maximum present size of that portion of the universe that we could, even in principle, observe. The number of protons located inside this Hubble radius is, from (2.33)

$$N \sim 10^{80}. \tag{2.39}$$

This is not very much less than the total number of baryons (protons and neutrons), or even the total number of baryons and electrons too. Note that N is the *square* of the ubiquitous number 10^{40}, a remarkable apparent coincidence which will be discussed in Chapter 4.

The parameters H, q, S and N together characterize the large scale structure of the universe. Their values are partly determined by the initial conditions that were built into the cosmos at its creation. In the coming sections we shall see how the structure of the universe is highly sensitive to the actual values assumed by these numbers.

3

The delicate balance

In the previous chapter it was shown that the gross structure of many of the familiar systems observed in nature is determined by a relatively small number of universal constants. Had these constants taken different numerical values from those observed, then these systems would differ correspondingly in their structure. What is specially interesting is that, in many cases, only a modest alteration of values would result in a drastic restructuring of the system concerned. Evidently the particular world organization that we perceive is possible only because of some delicate 'fine-tuning' of these values. This chapter and the next will survey some of the more striking examples.

3.1   Neutrinos

The most ubiquitous objects in the universe are neutrinos. Theory indicates that the big bang produced $\sim 10^9$ neutrinos for each proton and electron, and these now bathe the universe. However, neutrinos interact exceedingly weakly with ordinary matter. The Earth, for example, is almost completely transparent to them. Consequently the belief that the universe contains a huge background of cosmic neutrinos cannot be directly verified experimentally on Earth.

Because of the enormous numbers of neutrinos, the large scale structure of the universe is very sensitive to their properties. Until recently, it had generally been assumed that neutrinos are strictly massless, and hence travel at the speed of light. In 1980 a number of experiments seemed to indicate that this longstanding assumption might be in error. Preliminary results suggest that the neutrino could have a rest mass of about 5×10^{-35} kg, or about 5×10^{-5} of the mass of an electron.

Actually, the situation is a little more complicated because it is known that there exists more than one sort of neutrino. Particle physicists currently envisage three neutrino species, or 'flavours' (see Section 1.4). One of the surprises of the recent experiments is that a neutrino may, apparently, change its flavour in flight, oscillating rapidly in identity between, presumably, all three species. The oscillation phenomenon is closely related to the alleged existence of a non-zero mass for at least the electron-neutrino.

Although a mass of 5×10^{-35} kg is extremely small compared to all other known particles, the high density of neutrinos in the universe (about 10^9 m^{-3}) implies that the accumulated neutrino mass could outweigh all the stars. Indeed, had the neutrino mass turned out to be, say, 5×10^{-34} kg instead, then the gravitating power of the primeval background would have caused a drastic alteration in the expansion of the universe, possibly even halting it completely before now. It is then a remarkable thought that an apparently insignificant change in such a tiny mass would result in our living in a *contracting* rather than an expanding universe.

A slightly enhanced neutrino mass would have other consequences too. The present temperature of the neutrino background is about 2 K, which means that if they have an appreciable mass most cosmic neutrinos would now be non-relativistic, and in fact have speeds below that of the escape velocity of clusters of galaxies. They therefore tend to accumulate near the centres of these clusters and constitute a soup through which the galaxies plough in their orbital motions within the group, and in their internal rotations. Although the stars and gas do not interact directly with the neutrinos via the weak force to any appreciable extent, the neutrinos can still act on the galaxies gravitationally. The effect is to cause a viscous drag on the galactic motion. Calculations by a team of theoretical physicists of the University of Texas at Austin indicate that the observed galactic structure would be threatened by the neutrino soup if the neutrino mass were somewhat greater than the quoted value. Clearly even a small increase in the neutrino mass would cause severe disruption to the galactic structure.

It should be noted, however, that a very heavy neutrino mass (for example comparable to that of the proton) would lead to the suppression of the abundance of primeval neutrinos as a result of Boltzmann's theorem, which favours the distribution of energy among the lightest particles (see page 33). In this case the cosmic background of neutrinos would cease to be important.

Interest is also attached to the extreme weakness of the interaction between neutrinos and ordinary matter. This interaction, while minute, is nevertheless of great cosmological significance. During the hot primeval phase of the universe, before the first second had elapsed, the temperature was in excess of 10^{10} K, and the cosmological material contained abundant positrons (see Section 1.5). The interaction of neutrinos and antineutrinos with electrons, positrons, neutrons and protons brought about the following processes:

$$p + e^- \leftrightarrow n + \nu, \quad p + \bar{\nu} \leftrightarrow n + e^+,$$

enabling protons and neutrons to transmute into each other. So long as the rate of these reactions was appreciably greater than the rate of cosmic expansion, they would enable a thermodynamic equilibrium to be maintained between neutrons and protons.

As explained in Section 1.5, under equilibrium conditions the abundance ratio of neutrons to protons is determined by the Boltzmann factor $\exp(-\Delta mc^2/kT)$, where Δm is the amount by which the neutron mass exceeds the proton mass. As the universe expands the expansion rate steadily falls. The reaction rates also fall, because the universe cools and the particle densities drop. The reactions shown above become sluggish until they eventually drop below the expansion rate. When that happens, thermodynamic equilibrium is destroyed, and the abundance ratio remains frozen with the value it had at the temperature when equilibrium failed.

The temperature, T_F, at which the equilibrium failed is determined by equating the cosmic expansion rate $(\dot{a}/a)$ with the rate of the above reactions. In the next chapter it will be shown that, in the primeval universe, the expansion rate was

$$\dot{a}/a \simeq (8\pi G\rho/3c^2)^{\frac{1}{2}} \tag{3.1}$$

where ρ is the total energy density of the cosmological material, dominated at that epoch by radiation. We may therefore use Stefan's law for ρ, with an appropriate numerical factor to represent contributions from several species of radiation. This factor will be of order unity. Substituting explicitly for the radiation constant a, one obtains in place of (3.1)

$$\dot{a}/a \sim (Gk^4 T^4/\hbar^3 c^7)^{\frac{1}{2}}. \tag{3.2}$$

Turning to the neutron and proton transmutation rate, this will be determined by the strength of the weak interaction g_w, and the temperature T. Combining these quantities into an expression with units of $(\text{time})^{-1}$, one obtains

$$\text{reaction rate} \sim g_w{}^2 k^5 T^5/\hbar^7 c^6. \tag{3.3}$$

Equating (3.2) and (3.3) produces an expression for the freeze-out temperature

$$kT_F \sim G^{\frac{1}{6}} g_w{}^{-\frac{2}{3}} \hbar^{\frac{11}{6}} c^{\frac{7}{6}}. \tag{3.4}$$

It is at this stage that an extraordinary coincidence is discovered. First, it so happens that the neutron–proton mass difference is only a little greater than the electron mass:

$$\Delta m \simeq m_e. \tag{3.5}$$

Second, the strength of the weak interaction is, apparently accidently, related to the strength of gravity through the following numerical concurrence:

$$(Gm_e{}^2/\hbar c)^{\frac{1}{4}} \simeq g_w m_e{}^2 c/\hbar^3 \sim 10^{-11}. \tag{3.6}$$

When these two numerical accidents are used in Eq. (3.4), one obtains

$$kT_{\mathrm{F}} \simeq \Delta mc^2. \tag{3.7}$$

But the vital Boltzmann factor is $\exp(-\Delta mc^2/kT_{\mathrm{F}})$ which, by relation (3.7), has exponent of order unity. This factor determines the neutron/proton cosmic ratio, and the above analysis clearly implies that the neutron abundance will be an appreciable, but not overwhelming fraction of the total nuclear particle content of the universe. Detailed calculations give a value of about 10 per cent.

Had the factor $\Delta mc^2/kT_{\mathrm{F}}$ not happened, seemingly by chance, to come out at about unity, this ratio would have been either almost zero, or almost 100 per cent. The nuclear content of the universe is thus highly sensitive to what appears to be a random numerical accident involving quite distinct areas of physics.

What are the implications of the fortuitous arrangements, (3.6) and (3.7), of fundamental constants? If kT_{F} had been somewhat greater than Δmc^2 (or if Δm had been appreciably less than m_{e}), then the exponent of the Boltzmann factor would be small and the factor itself very close to unity. For example, if $kT_{\mathrm{F}} = 10\Delta mc^2$, then the neutron/proton ratio would be 0.9. This would have had a profound effect on the subsequent structure of the universe, for the following reason.

When the temperature fell below 10^9 K (below the photodisintegration temperature of the deuteron), the free neutrons combined rapidly with the free protons to form deuterium. The deuterium then went on to form helium:

$$n + p \to D$$

$$D + D \to \text{intermediate steps} \to He^4.$$

The He^4 contains equal numbers of protons and neutrons. Assuming all the available neutrons became incorporated into helium, all the cosmic hydrogen is due to the residue of unmatched protons, the existence of which is due to the excess abundance of protons over neutrons caused by the Boltzmann factor. If the Boltzmann factor were close to unity, there would be little hydrogen left over.

Hydrogen plays a vital role in the chemistry of the universe. Without hydrogen there would be no organic material and no water. Planets like the Earth, with large oceans, could not exist. More drastically, hydrogen is the fuel of all ordinary stable stars, such as the sun. Without this fuel stars could still form, but their appearance and behaviour would be quite different. In particular, stars made of helium would suffer much shorter lives before exploding or burning out. It follows that the existence of nature's most significant macroscopic structures – hydrogen-burning stars – depends on the accidental numerical relationships between the fundamental constants summarized in (3.6) and (3.7).

The existence of hydrogen seems all the more remarkable when one remembers that neutrons and protons are really composite particles differing only in their u, d quark content. The mass difference Δm is only about 10^{-3} of the proton mass, that is it represents a tiny correction. If the correction were only one-third of this value, then free neutrons would be unable to decay into protons, because they would not have enough mass to produce the required electron. Furthermore, if the neutron mass were only 0.998 of its actual value (that is if the u quark were very slightly heavier than the d quark) then free protons would decay into neutrons by positron emission: $p \rightarrow n + e^+ + v$. In that case there would probably be no atoms at all!

If kT_F were much less than Δmc^2, the neutron/proton ratio would be close to zero and there would be little helium in the universe. This would probably have few major consequences (though the presence of primeval helium in stars does affect their characteristics to a certain extent). However, there is another reason why the weak coupling constant g_w probably could not vary too much either way without profoundly affecting the chemical structure of the universe. This concerns one of nature's more important and spectacular processes – supernovae.

When a heavy star has exhausted its nuclear fuel, the core of the star becomes unstable against gravitational contraction. No longer able to generate the heat to sustain its internal pressure, the core starts to shrink under its own weight. Under certain circumstances the shrinkage becomes a catastrophic implosion, taking only a fraction of a second for the core to achieve nuclear densities.

The implosion releases enormous gravitational energy, much of which is transported outwards by neutrinos. Although ordinary stars are transparent to neutrinos, the highly compact core of the star is so dense that there is an appreciable impedance exerted on the outgoing neutrinos. It is believed that the pressure exerted by the flood of neutrinos can blast away the outer envelope of the star into space. Thus, the core

Fig. 9. Supernova in a spiral galaxy. The photograph records the explosion of a single star, which for a brief duration appreciably enhances the luminosity of the entire galaxy. The distended appearance of the image is due to overexposure.

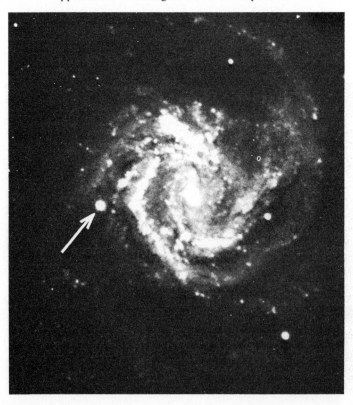

implodes and the periphery explodes. The explosion causes a huge increase in luminosity, so that for a few days the star can rival a whole galaxy in brightness. This titanic eruption is called a supernova and such eruptions occur about three times per galaxy per century.

Supernovae play an important part in the chemical evolution of galaxies. The galactic material of primeval origin is almost entirely hydrogen and helium. This raises the question of where all the other heavier elements came from. It is now known that they are synthesized inside stars. But how do they get out? The ageing star that explodes is rich in heavy elements that have been synthesized in its interior by successive nuclear reactions. The supernova explosion disperses this element-rich material around the galaxy. When subsequent generations of

Fig. 10. Supernova remnant. In 1054 oriental astronomers witnessed the explosion of a star in the constellation of Taurus. Today the debris appears as a ragged cloud of gas known as the Crab Nebula. Near the centre lies a rapidly rotating neutron star (pulsar), remnant of the collapsing core of the stricken star.

stars and planets form they incorporate the debris of these long-dead stars. We owe the presence of the carbon in our bodies, the iron core of our planet and the uranium in our nuclear reactors to supernovae that occurred before the solar system formed. Without supernovae, Earth-like planets would not exist.

If the weak interaction were much weaker, the neutrinos would not be able to exert enough pressure on the outer envelope of the star to cause the supernova explosion. On the other hand, if it were much stronger, the neutrinos would be trapped inside the core, and rendered impotent. Either way, the chemical organization of the universe would be very different.

3.2   **Nuclei**

In the previous section it was shown that the chemical structure of the universe is rather delicately dependent on the details of the weak nuclear force. Similar conclusions have been drawn about the strong nuclear force.

The strong force is responsible for binding atomic nuclei together against the electric repulsion of the protons. Being a short-ranged interaction, the strong force acts only between nearest-neighbour nuclear particles. In contrast the electric force operates between all the protons in the nuclei. It follows that a typical proton in a nucleus is glued in by the nuclear force of its near neighbours only, but is pushed by the accumulated electric field of all the other protons. In a large, heavy nucleus, with many neutrons and protons, the gluing force is no more powerful than in a light nucleus, but the electric force is greater due to the many protons. If the nucleus gets large enough, the electric force will actually exceed the nuclear attraction, and the nucleus will disintegrate.

In fact, nuclear instability sets in well before this extreme case. When a nucleus deforms in shape from its usual spherical configuration, the surface area increases. The additional particles which populate the extra surface are less strongly bound than when located in the body of the nucleus, because there are roughly half as many neighbours surrounding a surface particle as an interior particle. The nucleus therefore gains surface energy. However, the deformation reduces the electric energy by increasing the average separation of the protons. In light nuclei the surface energy gain outweighs the electric energy loss, and there is a tendency for the deformation to be squeezed away: the nucleus tries to adopt its lowest energy state. In heavy nuclei however, the electric energy dominates, and the deformation is amplified, leading to nuclear fission. Fission is exacerbated by quantum tunnelling effects which can lead to disintegration, with a lower probability, in still lighter nuclei. Asymmetric fission, such as the emission of alpha particles, is even more probable.

All known nuclei heavier than uranium have average lifetimes considerably shorter than the age of the Earth. If the strong nuclear force were somewhat weaker then there would be fewer stable chemical elements. It is hard to make quantitative estimates, because the nucleus is a complicated system, and the nuclear forces are still not properly understood, but it is probable that if the strong coupling constant g_s were, say, half its observed value, then nuclei such as iron, or even carbon, would be unlikely to survive for long.

More drastic consequences of changes in g_s follow from consideration of the simplest bound nucleus, the deuteron, consisting of a proton stuck to a neutron. The strong force that binds the deuteron has a range of about 10^{-15} m. To confine a particle within a range Δx implies, through Heisenberg's uncertainty principle, that the particle is unable to reduce its momentum below about $h/\Delta x$, corresponding to a kinetic energy of $h^2/2m(\Delta x)^2$. In the case of the deuteron this works out to be about 6×10^{-12} J. But the nuclear potential energy is only just greater than this. The net binding energy is 3.6×10^{-13} J. If the nuclear force were about 5 per cent weaker then the deuteron could not exist: the zero-point quantum energy would exceed the restraint of the nuclear attractive energy.

Deuterium plays a vital role in the power source of the sun and other stars. The sun burns through a sequence of nuclear reactions which starts with two protons fusing to form a deuteron, a positron and a neutrino

$$p + p \rightarrow D + e^+ + v.$$

This is a weak interaction process involving the inverse of beta decay, that is, a proton converts into a neutron. Further fusion involving deuterium now rapidly occurs:

$$D + p \rightarrow He^3$$
$$He^3 + He^3 \rightarrow He^4 + 2p$$

for example. This is a strong interaction process, merely involving the rearrangement of neutrons and protons, not their transmutation, and so it proceeds much more rapidly than the first reaction. Without deuterium, the main nuclear reaction chain used by the sun could not proceed. It is doubtful if stable, long-lived stars could exist at all.

Consequences still more profound would ensue if the strong interaction were only a few per cent stronger. It would then be possible for two protons to stick together. The di-proton is less stable than the deuteron for two reasons. First there is an electric repulsion between the protons. Second, the Pauli exclusion principle requires the protons to line up with their spins opposed, and the nuclear attraction is somewhat reduced in this configuration.

Once again, the observed situation is touch and go. The di-proton fails to be bound by a mere 1.5×10^{-14} J. This should be compared to an average binding energy of 1.3×10^{-12} J per nuclear particle in a typical nucleus. An increase in the nuclear coupling constant g_s by about 2 per cent would be enough to bind the di-proton. If that were the case, it would be energetically favourable for it to decay via the weak interaction to form a deuteron.

It was pointed out by Freeman Dyson that the existence of di-protons would render ordinary hydrogen catastrophically explosive. The sun burns its hydrogen slowly and steadily because the first link in the reaction chain shown above is controlled by the weak force, which permits only a very low rate of deuterium production. If deuterium could form via the di-proton, the important first step would then be under the control of the strong interaction, and would be about 10^{18} times more efficient. Catastrophic hydrogen consumption and energy release would then follow. Indeed, it is doubtful if any hydrogen would have survived beyond the hot primeval phase. The universe would be made almost entirely of helium, with consequences that have already been discussed in Section 3.1.

Nuclear structure and reactions also depend, of course, on the strength of electric forces. If the charge on the proton were larger, the stability of heavy nuclei would be threatened in the same way as if the strong nuclear force were weaker.

3.3   **Stars**

The essential features of stellar structure were outlined in Section 2.3, where it was shown that the energy density of radiation in a typical star is comparable with the kinetic energy of the particles. The structure of the star is, in fact, rather delicately dependent on the ability of the star to transport heat from its core by radiation. In the higher mass stars, as we have seen, radiation energy becomes dominant, and it is mainly by radiation flow that heat energy escapes from such stars. These are the so-called blue giants.

In somewhat lower mass stars this mechanism fails because the radiation cannot flow fast enough to keep the surface of the star sufficiently hot. This is important because unless the surface material remains partially ionized, instabilities occur which lead to the onset of convection. The convective turmoil supplements the radiative energy flow and prevents the temperature from dropping too far below the ionization temperature. Stars in which convection provides the dominant energy escape mechanism are therefore smaller and cooler than the blue giants, and are known as red dwarfs. The sun, and many other stable stars, lie in the rather narrow range delimited by the two extreme cases of blue giants and red dwarfs.

In Section 2.3 a formula for the mass of a typical star was derived: Eq. (2.18). What is remarkable is that this typical mass M_* just happens to lie in the narrow range between the blue giants and red dwarfs. This circumstance is in turn a consequence of an apparently accidental relation between the relative strengths of gravity and electromagnetism, as will now be shown. The treatment follows the original argument due to Brandon Carter.

The surface temperature T_s of the star is related to the central temperature T_c through the luminosity. The rate of radiant energy emission per unit area of the star's surface is $\frac{1}{4}acT_s^4$, so

$$acT_s^4 R^2 \sim L \sim 16\pi^2 \varepsilon^2 aT_c^4 R^4 m_e^2 c^5/Ne^4$$

using (2.25). Eliminating R by the use of (2.23) yields

$$T_s^4 \sim 16\pi^2 \varepsilon^2 T_c^2 m_e^2 m_p G^{\frac{1}{2}} h^{\frac{3}{2}} c^{\frac{11}{2}}/e^4 k^2. \tag{3.8}$$

The central temperature will automatically adjust itself so that the rate of production energy from nuclear burning is equal to the flow of energy from the surface. Nuclear reactions will switch on when the average thermal energy of a typical proton in the core approaches the energy required to penetrate the Coulomb (electric) barrier around other protons.

This energy is determined by two competing factors. The first is the energy distribution of protons in the core. This will be given by the Maxwell–Boltzmann law, which involves the factor $\exp(-E/kT)$. The numbers of protons with energy E much greater than kT (which are those most likely to penetrate the barrier) thus declines exponentially with E. On the other hand, the ease of barrier penetration rises with energy, because the protons are nearer to the top of the nuclear force barrier. There will be quantum tunnelling effects also assisting penetration. The resulting penetration factor is $\exp(-b/E^{\frac{1}{2}})$, where $b \simeq m_p^{\frac{1}{2}} e^2/4\pi\varepsilon\hbar$. The product of the penetration and Maxwell–Boltzmann factors peaks around $E = (bkT)^{\frac{2}{3}}$. It follows that the protons that are most effective in nuclear burning are those with energy close to this value. Prolific reactions will occur if this optimum value is not far from the average value; say $kT_c \sim 10^{-2}b^2 \simeq 10^{-2}m_p e^4/16\pi^2\varepsilon^2\hbar^2$. The temperature need not rise far above this to maintain a good supply of energy.

For the star to avoid convective instability, kT_s must exceed the ionization energy $\sim 0.1 e^4 m_e / 16\pi^2 \varepsilon^2 \hbar^2$, so from (3.8),

$$k^4 T_s^4 \sim 10^{-4} m_p^3 m_e^2 e^4 G^{\frac{1}{2}} c^{\frac{11}{2}} / 16\pi^2 \varepsilon^2 \hbar^{\frac{5}{2}} \gtrsim 10^{-4} e^{16} m_e^4 / (4\pi\varepsilon)^8 \hbar^8,$$

which reduces to

$$\alpha_G \gtrsim \alpha^{12} (m_e/m_p)^4, \tag{3.9}$$

where α is the electromagnetic fine structure constant. This remarkable relation compares the strength of gravity (on the left) with the strength of electromagnetism, and the ratio of electron to proton mass. Moreover, α is raised to the twelfth power, so the inequality is very sensitive to the value of e.

Putting in the numbers, one obtains 5.9×10^{-39} for the left hand side, and 2.0×10^{-39} for the right hand side. Nature has evidently picked the values of the fundamental constants in such a way that typical stars lie very close indeed to the boundary of convective instability. The fact that the two sides of the inequality (3.9) are such enormous numbers, and yet lie so close to one another, is truly astonishing. If gravity were *very* slightly weaker, or electromagnetism *very* slightly stronger, (or the electron slightly less massive relative to the proton), all stars would be red dwarfs. A correspondingly tiny change the other way, and they would all be blue giants. Carter has argued that a star's surface convection plays an important role in planetary formation, so that a world where gravity was very slightly less weak might have no planets. In either case, weaker or stronger, the nature of the universe would be radically different.

3.4 Galaxies

Astronomers still do not understand how galaxies formed, but clearly gravitational contraction must have played an important part. If the cosmic gases that emerged from the primeval phase of the universe were distributed more or less uniformly throughout space, then as the universe expanded, so the density would have steadily diminished. However, as explained in Section 2.4, here and there, where overdense regions of gas happened to exist, the gravitating power of the accumulated gas would have encouraged further accretion from the environment, thus enhancing the density perturbation. The gases in the vicinity of this blob would have experienced the competing tendencies of the cosmological expansion trying to disperse them, and the local gravitational influence of the blob attempting to restrain them.

Because of these opposing tendencies, the growth of the blob, and its eventual differentiation from the surroundings, would have proceeded only slowly. To adopt this scenario as a plausible model for galaxy formation, it is probably necessary to assume that the initial density irregularities were already quite pronounced.

In passing from a fairly uniform cloud of gas to a collection of compact stars, a great deal of gravitational energy must be dissipated. This can come about by the emission of heat from the cloud. As a blob of gas contracts, so it will heat up. The high temperature will prevent fragmentation into stars. However, eventually the gas cloud will radiate and start to cool. If the cooling rate is slow, star formation will still be inhibited. Probably the cloud would become disrupted and break into smaller blobs. On the other hand, if the cooling rate exceeds the contraction rate of the cloud, instabilities set in, and the cloud rapidly fragments into smaller and smaller subunits, the end product being individual stars.

According to this scenario, there is an upper limit to the size of a galaxy, determined by the competition between the contraction time, t_{shrink}, required for the cloud to shrink appreciably, and the cooling time t_{cool}, required for the cloud to cool appreciably. Gas clouds larger than this will not become galaxies of stars.

The contraction time is roughly the time required for a typical particle to fall from the edge to the centre of the blob under the gravitational force. For a cloud of mass M and radius R, this is, from elementary Newtonian theory

$$t_{shrink} \sim (GM/R^3)^{-\frac{1}{2}}. \tag{3.10}$$

The cooling time scale is a more complicated affair. If the gas is already at a low enough temperature, it can cool efficiently from processes such as radiation emission due to ionic recombination. In contrast, the cooling rate of a hot cloud will be much slower, arising mainly from radiation emission by free electrons (bremsstrahlung) near the surface. The theory of this process is straightforward but rather tedious and will not be reproduced here. As expected, it will depend on the Thomson scattering cross-section, (2.19), the mass m_e and density n, of electrons and the temperature T. Combining these quantities into an expression with the units of time, one obtains

$$t_{cool} \sim (1/n\alpha\sigma c)(kT/m_e c^2)^{\frac{1}{2}} \simeq (16\pi^2 \varepsilon^2 m_e^2 c^3/\alpha e^4 n)(kT/m_e c^2)^{\frac{1}{2}}.$$
$$(3.11)$$

At this stage one may note that the temperature of the cloud is related to its size. For diffuse gas, degeneracy pressure is absent, so Eq. (2.13) yields

$$kT \sim GMm_p/R. \qquad (3.12)$$

Although this relation was presented in Chapter 2 for the case of equilibrium, it is based on a fundamental theorem of mechanics (the virial theorem) and applies quite generally. In particular, it should be a good approximation for a gas cloud that has already shrunk appreciably. According to the central hypothesis, stars will be inhibited from forming in the cloud if $t_{cool} > t_{shrink}$. From (3.10)–(3.12) this condition reduces to

$$R > R_c \sim \alpha^4 \alpha_G^{-1} (m_p/m_e)^{\frac{1}{2}} a_0. \qquad (3.13)$$

Only if the ionized cloud could contract below the critical radius R_c would stars begin to form in abundance.

The size R_c is independent of the mass, and not very great (detailed calculations suggest R_c is not much greater than the size of the Milky Way). It follows that large, massive clouds will not readily form stars.

However, if the cloud is cool enough to be unionized, cooling is much more efficient, and stars will form easily. From (3.12) one sees that the low mass clouds are the cool ones, so there will be a critical mass, M_g, above which fragmentation of the cloud into stars will not occur, determined by the requirement that kT exceed the ionization energy for $R > R_c$. From (3.12) and (3.13) this restriction implies a *maximum* galactic mass

$$M_g \sim \alpha_G^{-2}\alpha^5(m_p/m_e)^{\frac{1}{2}}m_p. \tag{3.14}$$

How massive is M_g? Recalling from Eq. (2.18) that the mass of a typical star is $M_* \sim \alpha_G^{-\frac{3}{2}}m_p$, one obtains from (3.14)

$$M_g \sim \alpha^5\alpha_G^{-\frac{1}{2}}(m_p/m_e)^{\frac{1}{2}}M_*, \tag{3.15}$$

or about 10^{11}–10^{12} solar masses, which is a reasonable estimate. (The Milky Way has 10^{11} solar masses.)

In Section 2.4 an important cosmic parameter was introduced: the number of protons N in the observable universe. It was remarked that $N \sim 10^{80}$. Writing $N \sim \alpha_G^{-2}$, it then follows from (3.14) that the number of galaxies in the universe is

$$N_g \sim \alpha^{-5} \sim 10^{10} \tag{3.16}$$

where it is assumed that a typical galactic mass is $\sim (m_e/m_p)^{\frac{1}{2}}M_g$.

It is amusing to note that the number of stars in a typical galaxy is about the same as the number of galaxies in the universe, a coincidence which is now seen to follow, through (3.15) and (3.16), from the numerical accident

$$\alpha_G \sim \alpha^{20}. \tag{3.17}$$

4

Cosmic coincidences

The previous chapter reviewed a range of examples which provide convincing evidence that the nature of the physical world depends delicately on seemingly fortuitous cooperation between distinct branches of physics. In particular, accidental numerical relations between quantities as unconnected as the fine structure constants for gravity and electromagnetism, or between the strengths of nuclear forces and the thermodynamic condition of the primeval universe, suggest that many of the familiar systems that populate the universe are the result of exceedingly improbable coincidences.

Turning to the subject of cosmology – the study of the overall structure and evolution of the universe – we encounter further cosmic cooperation of such a wildly improbable nature, it becomes hard to resist the impression that some basic principle is at work. Many of the examples discussed in this chapter concern initial conditions on the universe rather than numerical relations. There is, however, one famous numerological case that provides the oldest example of the sort of coincidences that form the subject of this book.

4.1   The large numbers

In much of the analysis of the previous sections we have met the large number 10^{40}. Collecting together the examples:

$$\alpha_{\mathrm{G}}^{-1} \sim 10^{40} \tag{4.1}$$

$$N \sim 10^{80} = (10^{40})^2 \tag{4.2}$$

$$N_* \sim 10^{60} = (10^{40})^{\frac{3}{2}} \tag{4.3}$$

$$t_{\mathrm{H}}/t_{\mathrm{N}} \sim 10^{40} \tag{4.4}$$

$$t_N/t_P \sim 10^{40} \qquad\qquad (4.5)$$

or

$$t_H/t_P \sim 10^{60} = (10^{40})^{\frac{3}{2}}. \qquad\qquad (4.6)$$

The recurrence of this curious number in several apparently unconnected contexts has long been noticed by physicists and cosmologists.

Other, perhaps less striking, examples are

$$\alpha_w \equiv g_w m_e^2 c/\hbar^3 \sim (10^{40})^{\frac{1}{4}} \qquad\qquad (4.7)$$

$$S \sim (10^{40})^{\frac{1}{4}} \qquad\qquad (4.8)$$

$$N_{*_g} \sim N_g \sim (10^{40})^{\frac{1}{4}} \qquad\qquad (4.9)$$

where N_{*_g} and N_g are the number of stars in a galaxy and the number of galaxies in the universe, respectively.

Before embarking upon a discussion of the large numbers, a word should be said about the accuracy implied in the symbol $\sim$. Inspection of Table 5 shows that $\alpha_G^{-1} = 1.7 \times 10^{38}$, so that the use of the relation $\alpha_G^{-1} \sim 10^{40}$ might be regarded as somewhat straining the definition of an order of magnitude approximation. However, two points should be born in mind here. The first is that, compared to 10^{40}, even 10^2 is a minute fraction. Secondly, some of the factors that go to make up α_G are purely a matter of convention. For example, we could equally well have used h rather than $\hbar$. The choice in no way affects the general arguments presented here. The fastidious reader who wishes to rework the relations to greater precision is invited to use Tables 3 and 4.

In physical theory quantities like 4π or 3 frequently arise: their appearance occasions no surprise. Yet 10^{40}, which is constructed entirely from the fundamental constants of nature and therefore presumably has fundamental significance, is enormously greater than any of these more familiar quantities. The enormity of, for example, α_G^{-1} is an expression of the extreme weakness of gravity. Physicists have long wondered why gravity is so weak compared to the other forces of nature: compare, for example, electromagnetism with $\alpha^{-1} \simeq 137$. In recent years attempts have been made to unify the four fundamental forces of nature into a single mathematical theory. Early speculations suggested that a connection between gravity and electromagnetism might exist, and provide a numerical relation of the form

$$\pi ln\alpha_G^{-1} \sim \alpha^{-1}.$$

In 1967, Steven Weinberg and Abdus Salam presented a theory that combines the electromagnetic force with the weak nuclear force. In this theory the photon, which acts as the messenger of the electromagnetic force, is accompanied by other particles called W and Z, which transmit the weak

Table 5. *Useful data (SI units)*

Large numbers:	
$\alpha_G^{-1} \equiv hc/Gm_p^2 = 1.7 \times 10^{38}$	Inverse of gravitational 'fine structure' constant
$hc/Gm_e^2 = 5.7 \times 10^{44}$	
$hc/Gm_pm_e = 3.1 \times 10^{41}$	
$e^2/4\pi\varepsilon Gm_pm_e = 2.3 \times 10^{39}$	Ratio of forces in H^1 atom
Small distances (m)	
$h/m_pc = 1.3 \times 10^{-15}$	Compton wavelength of proton
$h/m_\pi c = 8.9 \times 10^{-15}$	Compton wavelength of pion
$e^2/4\pi\varepsilon m_ec^2 = 2.8 \times 10^{-15}$	Classical electron radius
Small times (s) $\sim t_N$	
$h/m_pc^2 = 4.4 \times 10^{-24}$	
$h/m_\pi c^2 = 3.0 \times 10^{-23}$	
$e^2/4\pi\varepsilon m_ec^3 = 9.3 \times 10^{-24}$	
Mass of pion	2.49×10^{-26}
Mass of neutron minus mass of proton	2.30×10^{-30}
Mass of Earth	5.98×10^{24}
Mass of Jupiter	1.90×10^{27}
Mass of sun	1.99×10^{30}
Mass of Milky Way galaxy	3.6×10^{41}
Mass of universe	10^{53}
Luminosity of sun	3.90×10^{26}
Radius of the Earth	6.37×10^6
Radius of sun	6.96×10^8
Binding energy of hydrogen atom	2.18×10^{-18}

nuclear force; W and Z are very massive particles, which accounts for the short range of the weak force (see Section 1.3). In the Weinberg–Salam theory, the link between the two forces is manifested in the following numerical relation

$$\alpha_w \sim \alpha(m_e/m_W)^2$$

where m_W is the mass of W. Note that the coincidence (4.7), in the form, $\alpha_w^4 \sim \alpha_G$ is the happy circumstance which, as pointed out in Section 3.1, leads to a universe with mainly hydrogen.

Following the success of the Weinberg–Salam theory in describing a variety of subatomic processes, and in reducing the total number of known forces from four to three, much effort has been devoted to a further unification of the resulting electroweak force with the strong nuclear force. Several schemes have been proposed. In these so-called grand unified theories, the photon and the W acquire yet another companion particle, this time exceedingly massive ($\gtrsim 10^{15} m_p$). One curious outcome of the mixing of the strong force, which couples to quarks, with the electroweak force, which couples to leptons (as well as quarks) is that the identities of quarks and leptons become blurred, enabling their transmutation to occur under some circumstances. For example, there is a tiny probability that the proton (which is made of quarks) will decay, ultimately into a positron (which is a lepton). The lifetime of the proton is predicted from this theory to be roughly $t_N(m_X/m_p)^4$, where m_X is the mass of the new superheavy mediating particle associated with the grand unification theory. To avoid conflict with experiment it is necessary that $m_X \gtrsim 10^{15} m_p$, leading to a proton lifetime $t_p \gtrsim 10^{30}$ years.

Most grand unified theories predict m_X in the range $(10^{15}$–$10^{16})m_p$, and more sensitive experiments are in progress to test the prediction of proton decay with an average life-time $\sim 10^{31}$ years. Note that if the prediction is confirmed $t_p/t_P \sim 10^{80}$, which is the square of the 'magic' number 10^{40}. The reason for this is that the quoted values of m_X come close to the Planck mass $m_P \equiv (\hbar c/G)^{\frac{1}{2}}$, for which $Gm_P^2/\hbar c = 1$. The Planck mass, formed entirely from the fundamental constants of gravity and quantum theory, is widely believed to play an essential role in any future theory of quantum gravity. The appearance of a mass m_X close to the value m_p suggests that a further synthesis – of the grand unified theory with gravity – may emerge from this approach. In that case there will be just a single, unified, fundamental force of nature, accounting for all interactions between matter. The apparent coincidence $m_X \sim m_p$ will then be explained.

Whether or not a fundamental physical reason can be found for the weakness of gravity based upon these ideas, it is a fact that if gravity were much stronger, the structure of the universe would be drastically altered; for example, it was mentioned in Section 3.3 that all stars would then be blue giants. Worse still, the entire universe would be unstable against gravitational collapse, and would probably have imploded before now. The Newtonian free fall time – the time for a spherical ball of matter to implode – is about $(GM/R^3)^{-\frac{1}{2}}$. Taking M as the mass of the observable universe (about 10^{53} kg) and R as its radius (10^{26} m), then the collapse time is around 10^{11} years (compare with the actual age of 2×10^{10} years). Evidently if G were only moderately increased, the universe would have disappeared by now.

Because 10^{40} is so unusually large, it is all the more striking to encounter this same number in several apparently different contexts. It was noted by Sir Arthur Eddington, and Paul Dirac, that the age of the universe in some natural atomic or nuclear units is also very close to 10^{40}. This fact is summarized in relation (4.4). There seems to be no obvious reason why the age of the universe should be related numerically to the number of particles in the universe, as implied by relations (4.2) and (4.4). Some physicists have been so impressed by this apparent coincidence of two such unlikely numbers that they have ascribed deep physical significance to it. Dirac wrote in 1938: 'Such a coincidence we may presume is due to some deep connexion in Nature between cosmology and atomic theory'.

There is, of course, a peculiarity. The age of the universe is not a fundamental constant, but changes with time. The quantity t_H is simply the epoch at which we happen to be living. Dirac therefore suggested that G should not be regarded as a constant of nature either, but should vary in proportion to $1/t$ so that the numerical coincidence

$$t_H/t_N \sim \alpha_G^{-1} \tag{4.10}$$

should remain true at all epochs. Theories of this type, involving time-dependent G, have frequently been studied over the years. Although a number of detailed theories which lead to the prediction of this phenomenon have been presented, no convincing observational evidence exists in support of a changing G (see Section 2.1).

An alternative explanation of the large number coincidence (4.10) has been proposed by Dicke. At the heart of his argument is the attempt to answer the following question. Is the fact that relation (4.10) happens to be true at the present epoch just the result of pure chance, or is there some special reason why we happen to be living at this particular epoch rather than any other? The issue clearly involves human observers in a rather basic way, and forms part of a pattern of arguments that attempt to relate the structure of the physical world to our own existence. Dicke's argument, and a number of similar analyses, will be discussed in detail in the next chapter.

The magic number 10^{40} crops up yet again in a quite different guise. This concerns the total number of charged particles in the universe, N, a concept introduced briefly in Section 2.4. Most of the cosmic particles are protons and electrons, so one arrives at the number N by dividing the mass of the universe by m_p and doubling it. The answer comes out at about 10^{80}, which is relation (4.2).

If this further coincidence is written in the form

$$N \sim (t_H/t_N)^2 \sim \alpha_G^{-1}(t_H/t_N) \sim 10^{40} \times 10^{40} \tag{4.11}$$

using (4.4), then rearranging and substituting for t_N the light travel time across the proton's Compton wavelength, one arrives at the important relation

$$G\rho t_H^2/c^2 \sim 1 \tag{4.12}$$

where $\rho = Nm_p c^2/(ct_H)^3$ is the average energy density of matter in the universe.

At this stage it is necessary to be more precise about what is meant by 'the universe'. In arriving at (4.12) we have taken the volume of the universe to be $(ct_H)^3$, which is the Hubble volume. The significance of this can only be appreciated by a digression into cosmological theory.

4.2   Cosmic dynamics

In Section 1.5 it was pointed out how the dynamics of a homogeneous isotropic universe can be described by a single parameter $a(t)$. The scale factor will obey an equation of motion determined by the laws of gravity. Curiously, it turns out that both Newtonian theory and general relativity yield the same differential equation

$$\dot{a}^2/a^2 c^2 + k/a^2 = 8\pi G\rho/3c^4. \qquad (4.13)$$

At our epoch, a is taken to be one. The parameter k, which should not be confused with Boltzmann's constant, has units of $(\text{length})^{-2}$. In the general relativistic theory it possesses a simple geometrical interpretation. The shape of space at any one instant need not be the flat space associated with Euclidean geometry. Einstein's theory predicts that, in general, space is warped, or curved. In a uniform universe this curvature must be everywhere constant.

Two possibilities arise. One is that of positive curvature, corresponding to $k > 0$. In this case space is closed, and finite in volume. The situation is analogous to the surface of a ball, which is finite in area, yet homogeneous (see Fig. 11). Nowhere does the balloon's surface have a boundary or edge, nor is there a centre. A positively curved universe would be a three-dimensional version of the surface of a ball, and would share with it the property of being circumnavigable. A projectile fired from any point would, after a sufficient duration, return to that point from the opposite direction.

Of course, the real universe is also expanding, so it would be more closely analogous to a balloon that is being inflated. This pictorial model enables ready visualization of the big bang.

When the balloon is deflated back to time zero, its area shrinks to nothing and it disappears. The creation event thus amounts to the sudden appearance of space, as well as matter. It is *not* the explosion of a lump of matter in a pre-existing void.

The case of negative curvature is easier to visualize, for space is infinite when $k < 0$. The curvature is outwards rather than inwards (like the ball). It is a geometry that is sometimes represented by a saddle-shaped surface in the case of a two-dimensional analogue (Fig. 11(ii)).

The special case $k = 0$ is when the spatial curvature vanishes: space is flat and infinite, and ordinary Euclidean geometry applies. Although the topology of the $k > 0$ universe differs from the cases $k \leq 0$, the local geometry varies smoothly as we consider a class of models with k passing

Fig. 11. The shape of the universe. According to Einstein's general theory of relativity, space can be curved by gravity. In a uniform cosmos only three shapes are possible. The drawings illustrate the three cases by analogy with two-dimensional surfaces to represent three-dimensional space. The dots represent galaxies, distributed more or less uniformly throughout space.
(i)   Space is curved positively into a finite volume (depicted here as a finite surface area). An adventurous astronaut could circumnavigate the universe.
(ii) Space is curved negatively. Only a portion is shown – the volume is infinite.
(iii) Space is flat and infinite, and the usual rules of Euclidean geometry apply.
In all three cases the surfaces should be envisaged as expanding.

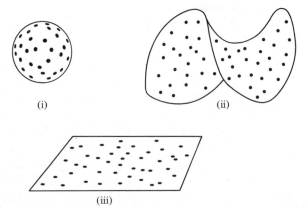

(i)   (ii)

(iii)

through zero. For k approaching zero from negative values, the space flattens out progressively until it becomes Euclidean at $k = 0$. For small positive values of k the curvature is small, and the (finite) volume of space is very large. As k increases, so the 'radius' of the universe, and hence its volume, shrinks. Standard geometry supplies an expression for the volume: $2\pi^2 k^{-\frac{3}{2}}$.

When $k > 0$ and space is finite, the notion of the total number of particles in the universe is well-defined, but what about the infinite cases, $k \lesssim 0$? At this stage it is necessary to take account of the causal structure of spacetime, as determined by the propagation of light signals in the distorted, expanding geometry.

Returning to the inflating balloon analogy, imagine a pulse of light emitted at the point A shown in Fig. 12. As the light travels across the universe, so the regions remote from A recede due to the expansion. The more distant regions recede faster. The light pulse has therefore to chase after the retreating galaxies. At any one time, there will be galaxies that have not yet received any light from point A, because the light that

Fig. 12. Horizon in space. Light emitted from galaxy A chases after the other receding galaxies (for example B). Two effects compete: the wavefront of light (broken line) expands outwards, and the universe expands. For distant galaxies, the light emitted at A in the big bang still has not reached them; the cosmic expansion is 'faster' than the expansion of the wavefront of light. Those galaxies that have not yet been overtaken by A's primeval light cannot, by symmetry, be seen by A. The limit of A's observable universe is called A's particle horizon. Because the cosmic expansion is decelerating, the wavefront eventually overtakes all galaxies, so the horizon encompasses more and more galaxies as time proceeds.

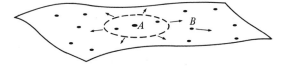

was emitted at the earliest moment – the big bang creation – has not yet caught up with them.

At first this result seems paradoxical, because in the beginning, corresponding to the phase when the sheet is very shrunken, space was compressed and all the galaxies were close together. It seems as though when space was shrunk to a very small volume, light must have been able to traverse it very rapidly. However, it is necessary to recall that at this primeval stage the universe was also expanding much more rapidly, so the light had to chase regions of the universe that were expanding very much faster than now. We therefore have to contend with competing limits. The distances for light to travel approach zero as we probe back to the creation event at $t = 0$, but the expansion rate rises to infinity. Which effect will win?

A simple analysis reveals that so long as the expansion rate decelerates from the first event at $t = 0$, then the light will not have been able to traverse the entire universe by now. As both the examples (1.16) and (1.17) involve an expansion rate that slows with time, this is almost certainly the case in the real universe.

If light from A has still not reached some galaxies, then those galaxies can know nothing of A. Because all physical influences must travel at the speed of light or less, no causal connection can exist between A and those galaxies. And the situation is symmetric: no influences from those galaxies can have reached A. It follows that at any one moment there exists around A a limiting distance, beyond which the universe can have no physical effect on A. Observers at A can see no events beyond that distance, except by waiting until the first light finally reaches them. This division into what can be seen and what cannot be seen at one time is reminiscent of the Earth's horizon, and is therefore referred to as the cosmic horizon, or more often, the particle horizon. It divides the universe around each place into the particles that can be seen from that place and those that cannot. The particle horizon expands at the speed of light, and in a decelerating universe will eventually overtake any galaxy, however far away it is.

Of course, if the expansion were to start accelerating again, this conclusion could not be drawn.

Particle horizons provide a natural definition of what is meant by the 'observable universe'. At any epoch t_H an observer can only see out as far as the horizon, which is the light travel distance since the creation, or about ct_H (the so-called Hubble radius). This is the definition used at the end of the previous section to compute the volume of the universe. Clearly it is proportional to time, as is the ratio t_H/t_N. On the other hand, consider N, the number of particles enclosed within the particle horizon at the epoch t_H. The horizon grows in proportion to time, so at first sight it appears as if N is proportional to (time)^3. However, the density of particles falls as the universe expands. Accepting relation (1.16), the density declines in proportion to (time)$^{-2}$. Thus N is only proportional to time, not (time)^3. Obviously, then, the coincidence $N \sim (t_H/t_N)^2$ will not be true at all epochs, because the left hand side grows linearly with time, whereas the right hand side grows quadratically.

It might be wondered whether, if $k > 0$, the coincidence (4.11) would occur if we chose for N the total number of particles in all of (finite) space, rather than only those that lie within the present horizon. It turns out that if k is fairly large the horizon encompasses an appreciable fraction of space at our epoch, so the two definitions of 'universe' are almost equivalent. However, we shall see below that $k \simeq 0$. For this reason we shall keep to the definition of N as the number of charged particles within the particle horizon.

Returning to Eq. (4.13), solutions are readily found in the rather special case that $k = 0$. For ordinary matter $\rho \propto a^{-3}$, because $\rho a^3/c^2$ is the mass of matter in the (expanding) volume a^3, and this remains constant in a homogeneous universe. Equation (4.13) then integrates to give $a \propto t^{\frac{2}{3}}$ as quoted in Section 1.5. For a universe dominated by radiation energy, $\rho \propto a^{-4}$. Integrating (4.13) with $k \simeq 0$ then yields $a \propto t^{\frac{1}{2}}$.

Although these solutions represent a special case, they provide a good approximation to the situation during the primeval epoch. This is because, when a is small, the term

k/a^2 is negligible compared to $\rho \propto a^{-4}$. We therefore expect $a \propto t^{\frac{1}{2}}$ to be a good description of the primeval phase.

For $k \neq 0$, there exists a characteristic epoch $(c^4k^2)^{-\frac{1}{4}}$ at which the k term begins to take effect, and can no longer be ignored. For $k < 0$, this term eventually dominates. When this is the case one may neglect the matter entirely. Putting $\rho = 0$ in (4.13) yields $a \propto t$. The universe expands at a uniform rate, without deceleration.

For $k > 0$, a more spectacular thing happens. The deceleration is enhanced by the k term, and eventually halts the expansion completely, at a time $\sim 1/ck^{\frac{1}{2}}$. Thereafter the universe begins to contract and eventually obliterates itself at a final crunch, like the big bang in reverse.

What determines the value of k? The quantity $\dot{a}/a$ is the reciprocal of the Hubble time t_{H}. If $k = 0$, Eq. (4.13) yields

$$8\pi G\rho t_{\mathrm{H}}^2/3c^2 = 1 \tag{4.14}$$

which is an exact version of relation (4.12). Equation (4.14) therefore supplies a characteristic critical energy density of matter

$$\rho_{\mathrm{crit}} = 3c^2/8\pi Gt_{\mathrm{H}}^2 \tag{4.15}$$

for which the universe is spatially flat, and expands like $t^{\frac{2}{3}}$ once matter rather than radiation becomes the dominant source of energy.

It follows from (4.13) that if $\rho > \rho_{\mathrm{crit}}$ then $k > 0$, the universe is spatially closed, and will eventually contract. The additional gravity of the extra-dense matter will drag the galaxies back on themselves. For $\rho < \rho_{\mathrm{crit}}$, the gravity of the cosmic matter is weaker and the universe 'escapes', expanding unchecked ($a \propto t$) in much the same way as a rapidly receding projectile. The geometry of the universe, and its ultimate fate, thus depend on the density of matter or, equivalently, on the total number of particles in the universe, N. We are now able to grasp the full significance of the coincidence (4.12). It states precisely that nature has chosen N to have a value very close to that required to yield a spatially flat universe, with $k = 0$ and $\rho = \rho_{\mathrm{crit}}$.

Inspection of Eq. (4.13) reveals that, because there is no evidence for the dominance of the curvature term k/a^2 at our epoch, it must still be small compared to the other terms in the equation. In particular

$$|k|/a^2 < 1/c^2 t_{\mathrm{H}}^2 \qquad (4.16)$$

where $|k|$ denotes the magnitude of k (which may be positive or negative). Taking $a = 1$ at our epoch, Eq. (4.16) implies that the radius of curvature of space, $r_{\mathrm{s}} \equiv |k|^{-\frac{1}{2}}$, is at least as great as the present Hubble radius ct_{H}.

It is amusing to consider what would have happened if N were, say, 10^{86} rather than 10^{80}. In that case the cosmological expansion would only have lasted about 10^8 years, and the universe would long since have collapsed out of existence. Similarly if N were 10^{77}, the k term would have taken control of the cosmic dynamics aeons ago. The more rapid expansion rate ($a \propto t$ rather than $t^{\frac{2}{3}}$) would have had a seriously inhibiting effect on the formation of galaxies. In either case, the structure of the universe would have been very different had the coincidence (4.12) not occurred.

Just how remarkable is relation (4.12)? Present observations indicate that $0.01 < \rho/\rho_{\mathrm{crit}} < 10$ so $(\rho - \rho_{\mathrm{crit}})/\rho_{\mathrm{crit}}$ lies anywhere between about -1 and $+9$. This does not, perhaps, seem so remarkable. However, one must remember that ρ is time-dependent. From (4.13) and (4.15) one obtains

$$(\rho - \rho_{\mathrm{crit}})/\rho_{\mathrm{crit}} = kc^2 t^2/a^2. \qquad (4.17)$$

Passing to early epochs of the universe, when radiation energy dominated the dynamics, $a \propto t^{\frac{1}{2}}$, so the above ratio is proportional to t. Thus, even if this ratio today differs from zero by, say, order unity, at one second after the creation it was a mere 10^{-18}. At the Planck time – the earliest epoch at which we can have any confidence in the theory – the ratio was at most an almost infinitesimal 10^{-60}. If one regards the Planck time as the initial moment when the subsequent cosmic dynamics were determined, it is necessary to suppose that nature chose ρ to differ from ρ_{crit} by no more than one part in 10^{60}.

We know of no physical reason why ρ is not a purely arbitrary number. Nature could apparently have chosen any value at all. To choose ρ so close to ρ_{crit}, fine-tuned to such stunning accuracy, is surely one of the great mysteries of cosmology? Had this exceedingly delicate tuning of values been even slightly upset, the subsequent structure of the universe would have been totally different. If the crucial ratio had been 10^{-57} rather than $< 10^{-60}$, the universe would not even exist, having collapsed to oblivion after just a few million years.

Why is ρ so close to ρ_{crit}? Put differently, why is k so close to zero? In the absence of any physical reason to explain the value of k, one is prompted to examine the fundamental parameters of the theory to determine whether they contain a characteristic curvature. One might then reasonably expect k to differ from this characteristic value by no more than a few orders of magnitude.

The parameters involved in the present theory are G, h and c. The characteristic curvature that can be built out of these quantities is $c^3/Gh \sim 10^{70}$ m^{-2}. This is some 10^{60} times greater than the actual value of k/a^2 at the Planck time. Had nature opted for the 'natural' value of k/a^2, the universe would have lived for only about $t_p \sim 10^{-43}$ s before either collapsing to nothing or exploding rapidly into emptiness. To achieve a universe with a longevity some 60 orders of magnitude longer than the natural fundamental unit of cosmic time t_p requires a balancing act between ρ and ρ_{crit} of staggering precision.

To summarize, in physical language what has happened is this. The energy density of matter in the universe, represented by ρ, determines its total gravitating power. A high density universe exerts more gravity, and causes a more rapid deceleration of the expansion. If the density is greater than the critical value, ρ_{crit}, then gravity beats the expansion and succeeds in reversing the cosmic motion to a catastrophic collapse. If ρ is very much greater than ρ_{crit}, then this reversal (followed by obliteration) occurs sooner. Conversely, if the density is very low, the gravitating power of the universe is small, and the expansion proceeds more or less unchecked. The lower the density, the more rapidly the expansion disperses the cosmic material. Unless ρ is exceedingly close to ρ_{crit}, the universe would either rapidly collapse back on itself, or explode.

The same balancing act can be viewed from the opposite point of view. For a given density of cosmic material, the universe has to explode from the creation event with a precisely defined degree of vigour to achieve its present structure. If the bang is too small, the cosmic material merely falls back again after a brief dispersal, and crunches itself to oblivion. On the other hand, if the bang is too big, the fragments get blasted completely apart at high speed, and soon become isolated, unable to clump together into galaxies. In reality, the bang that occurred was of such exquisitely defined strength that the outcome lies precisely on the boundary between these alternatives.

The situation is closely analogous to the motion of a projectile fired vertically from the Earth's surface. If it is fired too slow for the Earth's gravity, it will soon fall back to Earth. If it is fired too fast, it will shoot off into space and rapidly recede, never to return. The dividing line between these two is when the projectile is fired with exactly the so-called escape velocity – the minimum velocity for it to just escape the Earth's gravity. In the cosmological case, the expansion began with a strength so closely tuned to its gravitating power that it has just about 'escaped' its own gravity.

We finish this section by noting an alternative way of looking at this remarkable circumstance. For a radiation dominated universe the temperature $T \propto t^{-\frac{1}{2}}$. The energy kT supplies a natural unit of length, $\lambda = hc/kT$, which is the wavelength of a typical photon of the radiation. From (4.16) one obtains another unit of length: the radius of space curvature which, as remarked, is at least ct_{H}. Now (4.16) reveals that $|k| \propto (a/t_{\text{H}})^2 \propto t^{-1}$, for a radiation dominated universe, so the radius of space curvature $r_{\text{s}} \equiv |k|^{-\frac{1}{2}} \propto t^{\frac{1}{2}}$, which is precisely the same time-dependence as λ. The ratio

λ/r_s is therefore independent of time in a radiation dominated universe. The present value of λ corresponds to a typical wavelength of the microwave thermal background radiation at 3 K, and is around 10^{-3} m. The Hubble radius is $\sim 10^{26}$ m, so the ratio is about 10^{-28}. However this is an overestimate, as the universe has been matter, rather than radiation, dominated since 10^5 years. This has depressed the temperature by a factor of ~ 10, so a better representation of the ratio is

$$\lambda/r_s \lesssim 10^{-29}. \tag{4.18}$$

The fact that the right hand side of (4.18) is such a small number is an alternative expression of the fact that ρ is so close to ρ_{crit}. Being independent of time (at least while the universe was radiation dominated), this ratio has a rather fundamental character, and it might be supposed that it should be placed alongside the other cosmological parameters as a characteristic cosmic number. The fact that it is close to the value $(10^{40})^{-\frac{3}{4}}$ seems to indicate yet another large-number coincidence. However, the ratio (4.18) is not actually independent of the other cosmological parameters, as will now be shown. Let us assume k is so small that the term containing k in Eq. (4.13) may be ignored. With the assumption $\rho = \mathfrak{a}T^4$ (the Stefan–Boltzmann equation), corresponding to a radiation dominated universe, one readily integrates (4.13):

$$kT = (45\hbar^3 c^5/32\pi^3 G \mathcal{N})^{\frac{1}{4}} t^{-\frac{1}{2}}, \tag{4.19}$$

a result used to obtain Eq. (1.23). In arriving at (4.19) we have substituted for the radiation constant $\mathfrak{a}$, taking into account the fact that there will be several species of radiation present. This is represented by the weighting factor $\mathcal{N}$.

The amazing feature of (4.19) is that the coefficient of $t^{-\frac{1}{2}}$ is given entirely in terms of fundamental constants h, c and G. It does not depend on the initial conditions at all. (This is because we ignored k.) Assuming $\mathcal{N}$ is not too different from unity, and using (4.16)

$$\lambda/r_s \lesssim (t_P/t_H)^{\frac{1}{2}} \sim 10^{-30} \tag{4.20}$$

where we have put $a = 1$, $t = t_H$, to correspond to our epoch. We thus recover (4.18) as a special case of the big-number coincidence (4.6).

In summary, we can state the fundamental fine-tuning mystery of the universe as follows: why is the universe so much bigger than a typical wavelength of its primeval heat radiation?

The parameter k is some 60 powers of 10 less than the 'natural' value defined by the Planck length. It therefore seems reasonable to suspect that a hidden symmetry principle is at work, forcing k to be *exactly* zero. Indeed, some authors have tried to link the condition $k = 0$ with Mach's principle (page 129). Yet we know that k cannot be precisely zero or there could be no galaxies: the curvature of space on the relatively small scale of clusters of galaxies is definitely nonzero. Only on a cosmological scale does it average so close to zero. It is hard to imagine a principle that forces k to be on average so small, and yet still large enough locally to permit the formation of galaxies.

4.3 Cooperation without communication

It has been mentioned several times that the universe is remarkably uniform on a very large scale. Naturally, on the scale of galaxies there is a considerable degree of clumpiness in the distribution of matter, and a certain amount of spread in the motions. But on a scale of, say, 10^{24} m and above, the distribution is homogeneous and isotropic to a high degree.

One of the best tests for isotropy is the measurement of the microwave background radiation. Present observations place limits of one part in 10^4 on the temperature variation with orientation. Unless we are at a privileged spatial location in the universe, we must assume that this isotropy is present everywhere, which in itself implies homogeneity. It also implies that the cosmological expansion is homogeneous and isotropic.

Why is the universe so uniform that we need only consider

one degree of freedom, $a(t)$, to describe its global dynamics? One answer is to simply assume that it was created that way. This only says that the universe is the way it is because it was the way it was, so it hardly constitutes a proper explanation.

The mystery becomes all the more profound when account is taken of the particle horizon. In the previous section it was mentioned that the horizon grows with the speed of light. Extrapolating backwards in time to the primeval universe one finds that at the Planck time the horizon was only about 10^{-35} m (the Planck length) in radius, and encompassed about 10^{-8} kg of matter. Today, that volume of space has swollen to about 10^{-15} m^3.

The horizon divides regions of space that are causally connected. Regions that are outside each other's horizon cannot know what the other regions are doing. We have no reason to suppose, then, that the early universe should cooperate in its behaviour over lengths much greater than 10^{-35} m. In particular, there is no physical reason why the expansion rate should be matched between regions separated by more than 10^{-35} m. Yet if that were so we should expect the universe today either to be chaotic over length scales $\sim 10^{-5}$ m, or to show some evidence for the dissipation of this chaos during the epoch $t > t_\mathrm{p}$, when neighbouring regions came into causal contact.

The currently observed universe was causally divided by the horizon into at least 10^{80} separate regions at the Planck epoch, yet as has been emphasized, the cosmos presents a remarkably uniform aspect. And this remark extends to regions of the universe that *even now* are causally disconnected.

This may be illustrated by an analogy (Fig. 13). Imagine a ship A in mid-ocean. A sailor in the crow's nest sights another ship B that has just sailed over the horizon to the west. Simultaneously he spots a third ship C just above the horizon to the east. Although both B and C are within sight of A, B cannot see C, nor vice versa. They lie beyond each other's horizon.

Similar remarks apply to the cosmological horizon. We may

view distant galaxies on opposite sides of the sky and infer that, because of their proximity to our horizon, they lie beyond each other's horizon, and cannot 'see' each other. Yet the galaxies that lie in these causally disconnected regions of the universe – regions that have never been in any sort of physical contact – look remarkably similar. Moreover, the galaxies populate those disconnected regions at the same density and recede from their neighbours at the same rate. How is one to explain such an extraordinary degree of cooperation without communication?

It is hard to resist the impression of something – some influence capable of transcending spacetime and the confinements of relativistic causality – possessing an overview of the entire cosmos at the instant of its creation, and manipulating all the causally disconnected parts to go bang with almost exactly the same vigour at the same time, and yet not so exactly coordinated as to preclude the small scale, slight irregularities that eventually formed the galaxies, and us.

One possible explanation is to assume that the universe began in a very non-uniform condition, with turbulent and chaotic motion and a very uneven arrangement of matter and energy. Then, as the expansion progressed, so the primeval chaos dissipated, leaving the currently observed smooth pattern of motion and uniform distribution of matter. Several

Fig. 13. Ship's horizons. The look-out on *A* can see both *B* and *C*, although *B* cannot see *C*, nor vice versa. Similarly in cosmology, we can see distant galaxies that cannot be seen from each other. These galaxies have never been in any causal contact, and yet they look and behave identically.

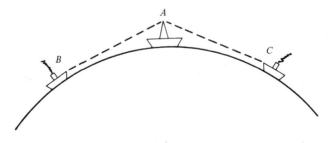

dissipation mechanisms suggest themselves: for example the conversion of turbulent gravitational energy into matter via the creation of particle–antiparticle pairs.

A considerable amount of effort has been expended investigating the theory of the dissipation of primeval chaos. One important process is undoubtedly the production of particles from the turbulent energy. But no convincing mechanism has yet been described to account for the present high degree of smoothness. It is always possible to find initial turbulence that cannot be damped completely away. Furthermore, some anisotropic motions tend to grow up again after the damping has ceased.

Another problem about the dissipation of primeval chaos is the excessive heat that it produces. All damping and frictional effects produce heat, and therefore entropy. Most of the cosmic heat resides in the microwave background radiation at 3 K, and this provides a constraint on the degree of dissipated turbulence in the early universe.

The essential difficulty can be illustrated by considering the presence of small amounts of anisotropy in an otherwise uniform universe. The expansion may still be approximately described by an average scale factor $a(t)$, but the gravitational equations will contain an extra term due to the more complex geometry. This is proportional to a^{-6}. We may rewrite (4.13) as

$$\dot{a}^2/a^2 c^2 + k/a^2 = 8\pi G\rho/3c^4 + A/a^6 \qquad (4.21)$$

where A is a constant and treat the anisotropy as a type of turbulent energy to go alongside the matter energy density ρ. As $\rho \propto a^{-4}$ it is clear that the anisotropy dominates the dynamics at early times (small a). Neglecting the k and ρ terms in (4.21) leads to the solution $a \propto t^{\frac{1}{3}}$ in contrast to $a \propto t^{\frac{1}{2}}$ found for the simple isotropic model.

The significance of the a^{-6} factor associated with the anisotropy energy is that it rises faster than the a^{-4} factor associated with the heat energy ρ as $a \to 0$. Hence, when the anisotropy is converted to heat, the amount of heat obtained for a given quantity of anisotropy is greater the earlier it was transformed. Stated differently, the anisotropy energy dies away faster than the heat energy as the universe expands. By converting to heat early on, the effect of the primeval anisotropy will be correspondingly greater.

How early on would the anisotropy realistically be dissipated to heat? The damping mechanisms so far examined are most efficient at the earliest times. But this is when the greatest heat is delivered for the least amount of anisotropy. If dissipation occurred at the Planck time (which is the preferred epoch for particle creation effects) then even an anisotropy of one part in 10^{40} would produce too much heat. Expressed another way, the present temperature of space requires that the expansion rate at the Planck time be fine-tuned in different directions to within one part in 10^{40}. This is another stunning example of 'cosmic conspiracy'.

The above argument depends on having a meaningful measure of cosmic heat (entropy). We measure the background heat energy density, but to construct a unit of heat requires a fundamental volume of space. Such a volume is provided by ordinary matter. The average density of protons in the universe is about one per cubic metre, so one cubic metre is a natural volume to choose.

Rather than discuss the heat energy per cubic metre, which varies with epoch as the universe expands, it is more appropriate to use the number of photons of heat radiation per proton, which is roughly independent of time. The photon/proton ratio S was introduced in Section 1.1; it has a value of about 10^9. It is the smallness of S (compared to 10^{40}) that constrains the initial anisotropy.

The parameter S is only meaningful so long as the number of protons in the universe remains fixed. It was mentioned in Section 4.1 that certain recent theories of the fundamental forces predict an unstable proton. If the proton number (strictly, the baryon number) is not a conserved quantity, then the photon/proton ratio is rendered meaningless. In model universes involving such effects, one can no longer appeal to the above reasoning to constrain the anisotropy using the 3 K background radiation.

4.4   The entropy of the universe

One of the fundamental cosmic parameters is S, the photon/proton ratio. The entropy density of heat radiation is proportional to the photon density, so S is also a measure of the entropy per proton in the universe. As explained in the previous section, this entropy could have been produced as the result of the dissipation of turbulence in the primeval universe, though it is very hard to understand why S is so small.

The total entropy of the universe is somewhat greater than that of the photons. Other species of radiation are likely to be present in the universe in addition to electromagnetic radiation: neutrinos, for example, and gravitons. We expect primeval neutrinos to bathe the universe, because before about 10^{-3} s neutrinos would have been coupled to matter, and hence to photons via the reactions discussed in Section 3.1. This would have ensured thermal equilibrium, maintaining the photons and neutrinos at a common temperature. After the neutrinos decoupled, the photon temperature would have been boosted a bit by the annihilation of muons and positrons. Calculations suggest that the present temperature of the neutrino background is about 2 K.

If there are three different neutrino species, this implies a neutrino entropy comparable to the photon entropy. A similar argument applies to gravitons. However, as the neutrino and graviton background radiation cannot be detected with foreseeable technology, we only have indirect evidence of its existence (see Section 3.1).

At first sight it may seem surprising that S is independent of time, because photons are continually being created and absorbed. In particular the emission of starlight enhances the photon content of the universe.

To investigate the question of accumulated starlight we may use the results of Sections 2.3 and 3.3. The number of stars in the observable universe is $\sim N/N_*$, each of average luminosity L given by $\sim \alpha c T_s^4 R^2$ and $kT_s \sim 0.1 e^4 m_e / 16\pi^2 \varepsilon^2 \hbar^2$, as explained in Section 3.3. The average lifetime of the stars is $t_* \sim t_H$ (see Eq. (2.30)), while the radius R is given by

$\sim \hbar^2/Gm_p{}^2 m_e N_*^{\frac{1}{3}}$, from the discussion in connection with Eq. (2.14). Combining all these factors and using (4.10), one obtains for the total number of starlight photons the expression

$$10^{-3}(m_e/m_p)\alpha^6 \alpha_G^{-\frac{1}{2}} N \sim 10N \qquad (4.22)$$

which should be compared with the number of primeval heat radiation photons $\sim SN \sim 10^9 N$. Clearly the starlight photons are far fewer. However, their energy is $\sim 10^4$ times greater, so that the accumulated starlight energy density is not very many orders of magnitude less than the primeval background energy density.

The ratio S has an important influence on the structure of the physical world. If S were 10^7 times greater, then the temperature of space would now be above the boiling point of water. No liquid water could exist in the universe until the cosmological expansion had reduced the background temperature appreciably. This would take several Hubble times, by which time most solar-type stars would have burned out (if they could have ever formed).

More importantly, even a modest increase in S would seriously threaten the existence of galaxies. As discussed in Section 3.4, galaxies formed by the growth of density perturbations in the primeval gases. This process could not begin until the gravitational dynamics of the cosmological medium were dominated by matter, rather than radiation. When did this happen?

The energy density of photons is $n_\gamma kT$, where n_γ is their number density. Similarly, the energy density of matter (mainly protons) is $n_p m_p c^2$. Equality occurs when $kT \simeq m_p c^2/S$. Using (4.19) to eliminate T we arrive at

$$t_{\text{equal}} \sim S^2 \alpha_G^{-\frac{1}{2}} t_N \sim 10^{13} \text{ s.} \qquad (4.23)$$

There is another criterion which must also be satisfied before galaxies can begin to grow. While the temperature of the universe was above the ionization temperature of hydrogen, the cosmic material was opaque to light, and was therefore subject to intense radiation pressure which would support it against rapid gravitational contraction. Once the temperature fell below about $0.1e^4 m_e/16\pi^2 \varepsilon^2 \hbar^2 k$, matter and radiation decoupled as the gases became transparent.

To compute the time of decoupling between matter and radiation, one notes that, after t_{equal}, the expansion scale factor follows the law (1.16). The temperature of the matter, however, falls at the same rate as the radiation temperature $(T \propto a^{-1} \propto t^{-\frac{2}{3}})$, so long as it remains ionized, and hence coupled to the radiation heat bath. Thus, assuming the decoupling epoch $t_{dec} > t_{equal}$,

$$T(t)/T(t_{equal}) = (t_{equal}/t)^{\frac{2}{3}}.$$

Using (4.23) for t_{equal} and $m_p c^2/kS$ for $T(t_{equal})$, one readily solves this equation for t. We require $kT(t) \sim 0.1e^4 m_e/16\pi^2 \varepsilon^2 \hbar^2$, for which t takes the value

$$t_{dec} \sim 10 S^{\frac{1}{2}} \alpha_G^{-\frac{1}{2}} \alpha^{-3} (m_p/m_e)^{\frac{3}{2}} t_N \sim 10^{13} \text{ s}. \tag{4.24}$$

This numerical coincidence, that $t_{equal} \sim t_{dec}$, has been a curiosity among astronomers for some time. It results from the numerical accident that the photon/proton ratio is

$$S \sim 10\alpha^{-2}(m_p/m_e). \tag{4.25}$$

Relation (4.23) is rather sensitive to the ratio S. Had S been below about 10^3, then t_{equal} would have dropped to a time, about 1 s, before primeval nucleosynthesis had begun. A universe dominated by matter rather than radiation expands at a different rate and will produce a greatly different ratio of hydrogen to helium. On the other hand, if S were greater than, say 10^{11}, then the universe would remain radiation dominated, and hence without galaxies, until the present epoch.

Evidently some rather basic features of our universe depend on S lying in the range of value $10^3 < S < 10^{11}$. But what determines the value of S?

After the cosmic background heat radiation was discovered in 1965, most astronomers and cosmologists supposed that the actual value of its temperature (3 K) was just an arbitrary number with no more special significance than the number of planets in the solar system. Some writers remarked that S lay close to the fourth root of the celebrated number 10^{40} (see Eq. (4.8)), but the uncertainty in its fairly modest value (compared to 10^{40}) rendered this apparent numerical co-incidence far less significant than the others already discussed.

An alternative to assuming that S is simply a reflection of initial conditions is that the entropy it represents was produced in some way during the primeval phase, due to certain dissipative processes. That is, the universe began with, say, $S \sim 1$, and the subsequent generation of heat boosted the value to 10^9. In that case, it is possible that the observed value of S can be computed from the details of the dissipative processes.

One example, anisotropy damping, has already been discussed in Section 4.3. Others include the dissipation of sound (the rumble of the big bang converted into heat), viscous effects and phase transitions between exotic superhot states of matter. One intriguing suggestion has been given by Martin Rees, who conjectures that a generation of large pregalactic stars may have formed and rapidly burned out (before t_{dec} and t_{equal}), their huge heat output being thermalized by the cosmic gases, which were still ionized. Putting the lifetime of a typical massive star (given by (2.28)) equal to t_{equal} (given by (4.23)) does indeed yield (4.8).

Recently a much more fundamental approach to the value of S has emerged. The mystery of why, in the big bang, nature created 10^9 photons for each proton is really only part of the greater mystery of how the protons themselves were created. In the laboratory protons are routinely produced in high energy collisions between subnuclear particles, but in every case the appearance of a proton is accompanied by an anti-proton (or a particle which rapidly decays into an antiproton). Individual protons can never be produced. Physicists have invented a quantity called baryon number to account for this. The proton, and many heavier particles, carry baryon number $+1$, the antiproton -1. A law of baryon number conservation then demands that each newly created proton must offset its baryon number by accompanying another particle with baryon number -1. Similar ideas apply to leptons: an electron must be accompanied by a positron. The simultaneous production of, say, just an electron and a proton, is forbidden by both baryon and lepton number conservation laws.

It seems reasonable to suppose that these conservation laws applied to the creation of matter in the big bang. In that case, for every proton produced there was a corresponding antiproton, and for every electron a corresponding positron. Now whenever a proton encounters an antiproton, or an electron meets a positron, annihilation occurs. Hence the primeval particles, being mixed together at high density with their antiparticles, would have had a very short existence. Nevertheless, in the high temperatures that prevailed in the big bang, fresh particle–antiparticle pairs would have been produced at a prodigious rate to offset the depletion from annihilation. Under the conditions of thermal equilibrium, which probably prevailed for most of the period from 10^{-35} s to 10^{-6} s, the opposing processes of pair creation and annihilation would have exactly balanced. It is then a straightforward matter to compute the equilibrium population abundances of the different particle species. Photons, being so readily convertible into particle–antiparticle pairs, would have been scarce. At 10^{-6} s, nearly all the particles would have been protons, neutrons, electrons, muons, pions and their antiparticles, with only a tiny fraction of photons.

Once the temperature fell below the value of $2m_p c^2/k$, the heat energy could no longer sustain the protons and antiprotons, so their rapid annihilation was no longer balanced by equally rapid replenishment. Later, by about 1 s, the remaining particles suffered a similar fate. Nearly all the rest mass energy of all these particles and antiparticles was converted into neutrinos and photons. Thus, the relative photon abundance abruptly shot upward.

Evidently a few protons and electrons escaped annihilation. If we assume that each proton–antiproton pair that annihilated gave of the order of one photon, then the present ratio $S \sim 10^9$ suggests that only one proton (and one electron) per billion escaped annihilation. The big question that this scenario then raises is: where are the antiprotons and the positrons that also escaped? If the laws of baryon and lepton conservation were respected, each particle in the universe should have, somewhere, a corresponding antiparticle.

Traditionally there have been two responses to this question. The first is to claim that the antiparticles are indeed extant: that the universe is an equal mixture of matter and antimatter. The most serious objection to this idea is that such a mixture is highly unstable, for any encounter between the two leads to explosive annihilation. If this were frequently happening in our galaxy it would produce a background of gamma radiation that is not observed. Estimates by Gary Steigman place an upper bound of only one part in 10^9 of our galaxy being composed of antimatter.

This problem could be circumvented by supposing that a large scale separation has occurred between the matter and antimatter, leading to the formation of whole galaxies consisting predominantly of one or the other. However, even galactic collisions occur from time to time, and the gamma ray data compels one to resort to separation on the scale of clusters of galaxies, rather than individual galaxies.

A matter–antimatter symmetric universe, whilst aesthetically appealing, begins to seem rather contrived, especially as no convincing mechanism has been proposed that can account for the separation of the two components into such large regions. The traditional alternative, however, seems equally unsatisfactory. This is simply to serenely accept that the universe is asymmetric – there are no substantial quantities of antimatter. The universe was made with a slight excess (about one part in 10^9) of matter, and it is this residue that has been left over from the big bang to form the galaxies – and us.

According to this scenario, the baryon and lepton content of our universe is simply an initial condition: something imprinted on the cosmos at its creation, and quite beyond rational explanation based on physical theory. In that case the photon/proton ratio S, which reflects this initial excess abundance of matter over antimatter, is a number chosen by nature at the outset, having the same status as, say, the ratio

m_p/m_e. We do not know why S has the particular value that it does, only that if it did not, the universe would be very different in its structure: once again, the unsatisfactory evasion that the world is what it is because it was what it was.

A much more persuasive idea is based on the recent attempts to unify the weak, electromagnetic and strong forces of nature into a so-called grand unified theory (GUT). These theories, as mentioned briefly in Section 4.1, predict that the proton can decay, ultimately into a positron. In so doing, the laws of baryon and lepton number conservation are both violated. The way is therefore open for the creation of matter without an equal quantity of antimatter.

The reason that these formerly sacrosanct laws are violated has to do with the nature of quarks and leptons. Many physicists regard these particles as truly fundamental, and it is therefore reasonable to enquire about the relation between them. There are probably six varieties of each. The quarks are subject to the strong interaction, while the leptons only feel the weak and electromagnetic interactions. In the unified theories, however, the distinction between these interactions, formerly considered quite separate, is blurred. Likewise the distinction between leptons and quarks is a little blurred, which is why a proton (made of quarks) can turn into a positron (a lepton).

At high temperatures, distinction between the three forces, and between leptons and quarks, is completely lost. Matter enters a new and rather featureless phase. The temperatures concerned are enormous, about 10^{28}–10^{30} K, at which the average thermal energy kT is not far short of the Planck mass-energy. These temperatures would have prevailed before about 10^{-35} s after the beginning of the big bang, and at that epoch the relationship between matter and antimatter would have been drastically altered.

Theory suggests that the single grand unified force that alone controlled all particle interactions (except gravity) was mediated by superheavy particles, playing a similar role to that played by photons, gluons, Ws and Zs. As the temperature fell, so the superheavies decayed into more familiar, lighter particles. But because of the possibility of baryon non-conservation, the decay products could have shown a slight preponderance of matter over antimatter, perhaps the required $1 + 10^{-9} : 1$ ratio which, following the eventual annihilation of all the antimatter, would leave $S \sim 10^9$.

What is so fascinating about this scenario is the chance of our being able to calculate the fundamental cosmic parameter S from basic physics – the physics of GUTs. A number of physicists have attempted this. Typically one obtains

$$S \sim (\text{ratios of quark masses}) \times (m_{\text{P}}/m_{\text{X}}) \qquad (4.26)$$

where m_{X} is the mass of the superheavy particles. Putting in the relevant numbers does indeed lead to estimates of S that are of the right order of magnitude.

If these ideas prove correct, it will mean that the important features of the universe which depend on $S \sim 10^9$, such as the existence of galaxies, and a hydrogen/helium ratio of about 4, are consequences of the values assumed by the fundamental parameters of the GUT, such as the superheavy masses. Had these differed by a fairly modest amount, the universe would be drastically different. In particular, if $m_{\text{X}} \ll m_{\text{P}}$, the universe would be extremely hot and without galaxies.

4.5 Cosmic repulsion

So far we have ignored the quantity Λ in the gravitational equations. As discussed in Chapter 1, there is no observational evidence that Λ differs from zero. As explained in Section 1.2, this so-called cosmological constant was originally introduced into physics by Einstein, who wished to construct a model universe which was static. (This was before Hubble's discovery of the expansion of the universe.) To do this, he had to propose a force that could counterbalance the attractive force of gravity between the stars, and this can be achieved with a Λ term in the gravitational field equations.

Unfortunately for Einstein, the static model universe that results, which happened to be spatially closed and finite in volume, is unstable. Slight perturbations would either cause it to collapse, or expand at an accelerating rate. Einstein dropped Λ in disgust. The subsequent discovery of the cosmological expansion removed the necessity for a static model of the universe, and with it the necessity for a Λ term. The influence of this unfortunate early experience with Λ led many cosmologists to believe on aesthetic grounds that it must be absent, that is that Λ must be strictly zero.

Modern developments have once again radically altered the picture. In Chapter 1 it was mentioned that Λ has the effect of a repulsive force acting across empty space. The concept of empty space is today a complex idea. According to the quantum theory emptiness – the absence of all particles – is not the same as inactivity: even a perfect vacuum is filled with forces and fields.

The Heisenberg uncertainty principle $\Delta E \Delta t \sim h$ permits an amount of energy, ΔE, to be 'borrowed' for a duration Δt. If Δt is short enough, this energy can be used to produce so-called virtual particles, which then rapidly disappear in order to repay the loan. This is the mechanism which explains how forces between source particles can be transmitted using a virtual 'messenger' particle, as discussed in Section 1.3. Virtual particles can appear, however, even in the absence of source particles. The vacuum state therefore contains limitless quantities of these short-lived particles, each existing only fleetingly, yet still able to interact and engage in complex processes. This seething, fluctuating *mêlée* exerts a gravitational influence in the same way as ordinary matter.

It is possible to compute the energy of this complex vacuum state. The calculations are tedious because of the appearance of infinite quantities that have to be circumvented to obtain meaningful results. The total vacuum energy will consist of contributions from all conceivable particles in nature – electrons, protons, photons, Ws, gluons, etc. Most of these contribute in a fairly straightforward way, but one important class of particles, called scalars, are more complicated. It is possible for the virtual scalar particles to form into a vacuum state in more than one way, and these alternative vacuum states differ greatly in their energy: in one typical case, equivalent to 10^{25} kg m^{-3}. It is a general physical principle that a system will seek out the state of lowest energy, and this is true of the vacuum. The higher energy states would be unstable and probably rapidly convert to the state of minimum energy.

One effect of this vacuum energy is to make a contribution to Λ. The dynamical behaviour of the vacuum is indistinguishable from a cosmological term in Einstein's gravitational field equations. So to the ordinary (sometimes called 'bare') Λ, we must add the quantum vacuum correction. The net quantity is what we actually observe in nature. And here is the astonishing feature. The quantum contributions to Λ are typically some *fifty* orders of magnitude greater than the maximum limit placed by observation on the actual value. Evidently the bare Λ and the quantum Λ are fine-tuned to almost exactly cancel each other to better than one part in 10^{50}. We must suppose that, to achieve a value of less than 10^{-53} m^{-2} (the observed upper limit), the vacuum contribution of, say, 10^{-2} m^{-2}, has the opposite sign to the bare Λ, and a magnitude that matches it to exceedingly great precision.

The size of the quantum contribution to Λ is determined by the microphysical parameters that enter into the particular field theory being considered. For example, the scalar particles in the Weinberg–Salam theory of the electroweak force contribute

$$\Lambda_q = -\pi G m_\phi{}^2/\sqrt{(2)}c^4 g_w \sim -10^{-2} \text{ m}^{-2},$$

where m_ϕ is the mass of the scalar particles. Yet

$$\Lambda_{bare} + \Lambda_q \lesssim 10^{-53} \text{ m}^{-2}.$$

If G, or g_w, differed from their actual values by even one part in 10^{50}, the precise balance against Λ_{bare} would be upset, and the structure of the universe would be drastically altered. Moreover, in the case of the so-called grand unified theories, the precision of the matching must be increased to better than one part in 10^{100}.

What effect would a value of Λ larger than 10^{-53} m^{-2} have? Recalling that it leads to a force which opposes gravity, and grows in strength with distance, the existence of a Λ term, however small, in an ever-expanding universe, implies that eventually the dilution of matter will be great enough for the repulsive force to gain ascendency. After that happens the universe will start to expand faster and faster.

The presence of the Λ term modifies Eq. (4.13) as follows:

$$\dot{a}^2/a^2c^2 + k/a^2 - \Lambda = 8\pi G\rho/3c^4. \tag{4.27}$$

Clearly, for small a, Λ may be neglected, as we have indeed done in the foregoing discussions about the primeval universe. However, for large a the effects of Λ dominate the other terms. Neglecting the k and ρ terms, Eq. (4.27) has the solution

$$a \propto \exp(c\Lambda^{\frac{1}{2}}t), \tag{4.28}$$

that is, the universe swells exponentially. The curve shown in Fig. 3 would then no longer apply at late times. Instead of curving continually downwards, the curve would turn up again at an accelerating rate.

In the case $\Lambda < 0$, the Λ force actually assists gravity so the universe contracts, on a time scale $\sim 1/c\Lambda^{\frac{1}{2}}$.

It is the absence of any sign of either exponential growth or recontraction that enables astronomers to place a limit on the magnitude of Λ. If Λ were a factor of 10 greater, then already the pattern of expansion of the universe would be quite different. But if Λ were several orders of magnitude greater, the expansion of the universe would be explosive, and it is doubtful if galaxies could ever have formed against such a disruptive force. If Λ were negative, the explosion would be replaced by a catastrophic collapse of the universe. It is truly extraordinary that such dramatic effects would result from changes in the strength of either gravity, or the weak force, of less than one part in 10^{40}.

It might be supposed that fine-tuning between Λ_q and Λ_{bare} to such stunning precision suggests a new fundamental principle at work which requires that Λ is precisely zero, as Einstein wished. Rather than attribute the smallness of Λ to an accidental cancellation of huge quantities, it could be regarded as a basic principle of physics, to be forced on nature.

Although the cosmological constant is indistinguishable

from zero at the present epoch, recent work on GUTs suggests that during the high temperature primeval phase, the vacuum energy could have been drastically different, temporarily giving Λ an enormous value. Some cosmologists believe that this repulsive force was actually the cause of the big bang. They suggest that at about 10^{-35} s the universe embarked on a brief but significant phase of runaway expansion, as Λ dominated over ordinary gravitational effects to produce a short-lived de Sitter phase. This model is known as the 'inflationary universe' scenario. It is claimed that inflation could provide a natural explanation for the homogeneity and isotropy of the universe, and for why ρ is so close to ρ_{crit}.

5

The anthropic principle

The catalogue of extraordinary physical coincidences and apparently accidental cooperation surveyed in the previous two chapters offer compelling evidence that something is 'going on'. At the beginning of Chapter 4 the comment was made that a hidden principle seems to be at work, organizing the cosmos in a coherent way. How else could one explain how the expansion energy of the universe is not only matched to its gravitating power to ensure survival for at least 10^{60} times longer than its natural cycle time, but is so matched in precisely the same way everywhere, even in causally disconnected regions of space? What other explanations would account for the almost exact cancellation between Λ_q and Λ_{bare}, or the complete absence of observable anisotropy?

In spite of the obvious need to discover a cosmic principle from fundamental physics that can explain these 'miraculous' characteristics, as well as the other astonishing coincidences from microscopic physics discussed in Chapter 3, no such principle has been proposed. Instead, the only systematic attempt to explain scientifically the seemingly contrived structure of the physical world rests not on fundamental physics at all, but on biology. This pattern of arguments appeals to an undeniable, yet at first sight irrelevant, feature of the universe – us.

Normally 'the observer' is discounted in consideration of physical science. We are here, it is usually assumed, just 'for the ride'. Some scientists have challenged this traditional assumption, declaring that the structure of the physical world is inseparable from the inhabitants that observe it, in a very fundamental sense. They argue that there does indeed exist a guiding principle which works to fine-tune the cosmos to incredible accuracy. It is not a physical principle, however, but an *anthropic* principle.

5.1   Implications for biology

The previous chapters will have convinced the reader that the structure of the physical world is delicately dependent on a variety of apparent numerical accidents. Many of the rather basic features of the universe are determined in essence by the values that are assigned to the fundamental constants of nature, such as G, α, m_p and so on, and these features would be drastically altered if the constants assumed even moderately different values. It is clear that for nature to produce a cosmos even remotely resembling our own, many apparently unconnected branches of physics have to cooperate to a remarkable degree.

All this prompts the question of why, from the infinite range of possible values that nature could have selected for the fundamental constants, and from the infinite variety of initial conditions that could have characterized the primeval universe, the actual values and conditions conspire to produce the particular range of very special features that we observe. For clearly the universe is a very special place: exceedingly uniform on a large scale, yet not so precisely uniform that galaxies could not form; extremely low entropy per proton and hence cool enough for chemistry to happen; almost zero cosmic repulsion and an expansion rate tuned to the energy content to unbelievable accuracy; values for the strengths of its forces that permit nuclei to exist, yet do not burn up all the cosmic hydrogen, and many more apparent accidents of fortune.

These coincidences have been noticed by many scientists. Both Eddington and Dirac were greatly impressed by the appearance of the number 10^{40} in different contexts (see Section 4.1), and built elaborate physical theories around them. Eddington attempted to derive the number 10^{40} from new fundamental physical principles, while Dirac proposed that this large number is actually time dependent, implying $G \propto t^{-1}$. Dirac's theory was later elaborated by Pascual

Jordan (and then by Dirac himself). Jordan also attempted to explain relation (4.3) on the basis of radical departures from conventional physics. In more recent years Carter has drawn attention to what he calls the 'remarkable coincidence' of relation (3.9) which ensures that typical stars lie between the blue giant and red dwarf conditions.

Many of these writers have been most impressed by the fact that our own existence as conscious organisms is delicately dependent upon the structure of the physical world that we perceive. They have argued that had any of the finely-tuned conditions discussed in the previous chapters failed, then life – certainly life as we know it – would have been impossible.

Carter remarks 'The existence of any organism describable as an observer will only be possible for certain restricted combinations of the parameters'. John Barrow states the issue somewhat differently: 'Our existence imposes a stringent selection effect upon the type of Universe we could expect to observe'. In a paper that begins with the memorable sentence 'We exist', D.V. Nanopoulos adopts the same sentiments on the issue of the ratio of photons to protons in the universe; 'Our existence', he maintains, 'puts strong limits on the ratio'.

Carter's comment is essentially that had the fundamental parameters assumed appreciably different numerical values we would not be here to comment on the fact. But Barrow and Nanopoulos seem to be saying something a little more positive: that our existence *constrains* the structure of the universe, indeed, it even 'selects' it – an idea captured by the words of John Wheeler: 'Here is man, so what must the universe be?'

Bryce DeWitt could be construed as expressing support for either of the two alternatives in a deliberately ambiguous and terse phrase: 'The world we live in is the world we *live* in'.

Invoking the human connection in this sweeping context has been dubbed the *anthropic principle*. Over the years it has come to mean many things to many people. Before discussing the various interpretations of the anthropic principle, it is helpful to begin with an illustration of the principle at its best, where the relevance of our existence cannot be denied.

5.2 Explaining the large-number coincidences

One of the earliest specific demonstrations that biology can be used to explain an otherwise mysterious feature of the physical universe is due to Robert Dicke. In 1961 he declared that Eddington and Dirac had been misguided in searching for new fundamental principles of physics to explain the apparent coincidence (4.10): that the age of the universe in nuclear units is the same huge order of magnitude as the ratio of electrical to gravitational forces between two protons.

Obviously, the *present* age of the universe is defined by the existence of the human community. Technological society, and with it the measurement of the fundamental constants under discussion here, has occupied a minute fraction of the life span of the universe, so can be regarded as defining a characteristic epoch t_{now}. The mystery is, why does t_{now} bear the same numerical relation to t_N as the electric force between protons bears to their gravitational attraction?

Dicke reasoned that t_{now} is not a randomly selected instant of time, but intimately connected with the time scales of certain physical processes in the universe that are themselves pre-requisites for the existence of intelligent life, and hence technology. One could imagine a variety of such pre-requisites, but the one chosen by Dicke concerns the existence of elements heavier than hydrogen. Life on Earth is based on the element carbon, while nitrogen and oxygen are also vital. These elements did not exist in the primeval universe. Their presence in reasonable abundance is attributed to the nucleosynthesis which occurs inside stars.

During the big bang, temperatures high enough to synthesize heavy elements were available, but only for a duration of a few minutes. Only the element helium (irrelevant for life) was produced in abundance. On the other hand, in stellar interiors, temperatures $\sim 10^7$ K or higher are available for billions of years, enabling a large fraction of the stellar material to become converted into heavy elements. In order for these elements to become the chemical building blocks of life, they must be dispersed around the galaxy. As explained in Section 3.1, this can occur when a star reaches the end of its life, and has exhausted its nuclear fuel. If the star is rather massive, then it is likely to explode violently as a supernova, spewing its contents into interstellar space. Our bodies are formed, as Sir James Jeans once remarked, from the ashes of long dead stars.

According to Dicke's reasoning, life cannot form in the universe until at least one generation of stars has passed through this life cycle, and seeded the galaxy with the supernova debris containing carbon. On the other hand, the consumption of hydrogen fuel by stars is irreversible, so that this cycle cannot be repeated *ad infinitum*. After a few generations of stars, the galaxy's supply of nuclear fuel will become severely depleted, and new stars (at least stable stars like the sun) will become rather rare. The galaxy will then begin to cool, and one imagines life will become impossible.

These ideas may lead one to suppose that life will only be found in the universe during the epoch from t_* to, say, $10t_*$, where t_* is the average life of a moderately large star. This quantity was computed in an approximate way in Section 2.3. From Eq. (2.28) we have

$$t_* \sim \alpha_G^{-1} t_N \sim 10^{40} t_N.$$

If one now identifies t_{now} with t_*, to within an order of magnitude, based on the reasoning that we, as living creatures, can only find ourselves perceiving the universe during the epoch from t_* to $10t_*$, then one of the large-number 'coincidences' is explained. It is seen not to be a coincidence at all, nor a manifestation of hitherto unknown physics, but a straightforward consequence of basic physics and biology.

Evidently a *biological* explanation of a fundamental feature of our world has succeeded where theoretical *physics* has failed. The fact that t_{now} is defined only to within an order of magnitude by t_* is not a serious shortcoming of this argument, as the big numbers are only approximately equal anyway, and can be defined to have a variety of values depending on whether, for example, one uses the electron rather than the proton mass. However, these differences are all very small compared to the huge magnitude of the number 10^{40}. The fact that t_* lies so close, in units of t_N, to this number, is striking.

It is possible to recover the lower limit on t_* by an alternative argument that makes no reference to the stellar evolution time scale. Bernard Carr and Martin Rees have pointed out that life depends on the existence of galaxies, and these can only form after the epochs t_{dec} and t_{equal} (see Section 4.4). If one accepts the GUT theory that *predicts* a photon/proton ratio $S \sim 10^{10} \sim \alpha_G^{-\frac{1}{4}}$ on the basis of fundamental physics (rather then regarding S as an initial condition and hence a free parameter) then Eq. (4.23) immediately yields

$$t_{equal} \sim \alpha_G^{-1} t_N$$

and hence $t_{now} \gtrsim t_{equal} \sim 10^{40} t_N$.

This line of reasoning – that human observers select a location in the space that may be atypical – is contrary to the spirit of the Copernican revolution. Nicholaus Copernicus, by denying the special status of the Earth in cosmic dynamics, initiated a tradition that has influenced scientific thinking for four centuries. In most respects, the Earth can be regarded as most unexceptional in status, and typical of a vast number of similar planets near similar stars in similar galaxies. Yet our existence as biological organisms has selected for us a location in space that is in some sense atypical. Although the Earth may not claim special status among planets, the fact that we find ourselves living on a solid surface, when the vast majority of the material in the universe is in the form of tenuous gas clouds or balls of hot plasma, and the fact that we are located near a stable star, when many stars have erratic behaviour or are grouped in multiple systems unsuitable for equable planets, is no coincidence. We could not, presumably, survive in the hostile environment associated with more typical cosmic material. Similarly, our temporal location in the cosmos is constrained by the fact that the universe *evolves*, and during its evolution from a hot, dense furnace to a collection of burnt-out, dispersed galaxies, only a relatively restricted time interval is suitable for life.

The Dicke argument can, of course, be faulted on a number

of grounds. First it is based on a conception of life as we know it. It is conceivable, but by no means probable, that extra-terrestrial life forms might exist that are based on wildly different physical processes. It is excessively chauvinistic, especially when appraising fundamental issues, to regard carbon-based chemical life as the only vehicle for intelligence and technology.

On the other hand there may be fundamental arguments as to why life of any sort cannot evolve to the level of pos-sessing intelligence until certain physical processes have been completed. Life, by any definition, involves a high degree of complexity and order which have certain pre-requisites. For example, the second law of thermodynamics, which regulates all natural activity, demands that a disequilibrium of some sort be present before order can arise. There may well be very basic principles that constrain the rate of accumulation of order and information, and thereby the rate of evolution of life, based on the existing forces of nature.

Perhaps the weakest point in the anthropic argument as used in the present example concerns the upper limit on t_{now}. Even though the galaxy will eventually become unsuitable for life as the stars burn out, in the billions of years left it would be surprising if technology did not advance to the point of overcoming this. One can easily envisage artificial environments suitable for life extending arbitrarily far into the future. Dyson has carried out a searching analysis of survivability and concludes that a sufficiently resourceful community could achieve unlimited longevity by careful manipulation of the environment. To overcome the inevitable decay of sources of free energy, and thereby avoid participat-ing in the famous 'heat death' assured for the cosmos on the basis of the second law of thermodynamics, it would be necessary for such a community to 'hibernate' for periods of increasing duration. Nevertheless the integrated life span of the community could still be infinite.

If life will indeed remain extant (if dormant) for the infinite future, it might be regarded as surprising that we find ourselves perceiving a universe of finite age. Certainly the large-number coincidence (4.10) would only have a lower limit: because t_{now} has no upper bound, the left hand side of (4.10) is only restricted by t_* to be greater than about 10^{40}.

This weaker restriction only applies in a universe that will continue to expand for ever. If the density of cosmic material is great enough, recollapse will occur, putting an end to all life. The fact that our t_{now} is then not too much greater than t_* would be explained if one could show that the total lifetime of the universe were not too much longer than t_*. This might indeed be the case if present experiments on neutrinos confirm the existence of a non-zero rest mass, (see Section 3.1). However, it is important to note that we know of no fundamental reason why the cycle time of a recollapsing universe should be related to t_*. If the cosmic lifetime turns out to be $\sim 10t_*$, this must be regarded as purely accidental: the former quantity depends on the initial expansion rate of the universe. Possibly a more detailed analysis will reveal that stable galaxies and stars can only form if ρ exceeds ρ_{crit} by just the amount necessary to yield a cosmic lifetime $\sim 10t_*$.

The central role of the element carbon in terrestrial life prompted Fred Hoyle to draw attention to a further curious accident of nature. Carbon nuclei are synthesized in stars as a result of the almost simultaneous encounter of three helium nuclei. Such a triple collision is, of course, rather rare, and would be utterly insignificant if it were not for a fortuitous property of the carbon nucleus. The union of two helium nuclei forms an unstable nucleus of beryllium, Be^8. The probability of the further incorporation of a third helium nucleus, to form carbon (C^{12}), before the decay of Be^8, depends sensitively on the energy with which the helium nucleus strikes the temporarily existing Be^8. The reason for this concerns the existence of so-called nuclear resonances. Roughly speaking, when the frequency of the quantum wave associated with the incoming helium nucleus matches an internal vibration frequency of the composite system, the nuclear cross-section for capture of the third helium nucleus rises very sharply. By chance, the thermal energy of the nuclear constituents in a typical star lies almost exactly at the location of a resonance in C^{12}. This happy accident ensures the efficient production of carbon inside stars. Without it, the rate of carbon formation would be very much reduced.

This is, however, only half the story, for it is necessary that the newly synthesized carbon survive the subsequent nuclear activity inside the star. Carbon will be depleted as it burns to form still heavier elements. Specifically, the further collision of a helium nucleus with C^{12} produces oxygen, O^{16}. Once more, though, nature has made a fortunate choice. A resonance in the O^{16} nucleus lies safely below the thermal energy of the constituents, so the C^{12} is spared the fate of being burned out of existence to form oxygen.

The details of nuclear structure are immensely complicated, but ultimately the location of the nuclear resonances depends upon the fundamental forces of nature, especially the strong nuclear force and the electromagnetic force. Had the strengths of these forces not been rather precisely chosen, the fortuitous arrangement of resonances in C^{12} and O^{16} would not have occurred and life, at least of the terrestrial variety, would have been exceedingly less likely.

Returning to this topic in a recent publication, Hoyle considers the carbon–oxygen synthesis coincidence so remarkable that it seems like a 'put-up job'. Regarding the delicate positioning of the nuclear reasonances, he comments: 'If you wanted to produce carbon and oxygen in roughly equal quantities by stellar nucleosynthesis, these are the two levels you would have to fix, and your fixing would have to be just about where these levels are actually found to be . . . . A commonsense interpretation of the facts suggests that a superintellect has monkeyed with physics, as well as chemistry and biology, and that there are no blind forces worth speaking about in nature'.

5.3   Weak and strong anthropic principles

Although both Dicke and Hoyle invoke the carbon connection in their discussion of superficially unlikely coincidences in nature, there is a distinct difference in the status of the two arguments as presented in the previous section.

In the case of the large numbers considered by Dicke, our existence as carbonaceous observers *explains* the concurrence of the number 10^{40} in two different contexts. Human life has *selected* an epoch from all those available, which is necessarily of order t_*, and hence such as to satisfy the big-number coincidence (4.10).

On the other hand, Hoyle's example, in the form stated above, does not *explain* the coincidence of nuclear energies, but merely comments on the extreme fortune of the circumstance: had it not been so we should not have been here to discuss the issue. It is one illustration of how, in a universe the structure of which is so delicately dependent on the constants of nature, we are exceptionally fortunate to exist.

The former line of reasoning has been called the *weak* anthropic principle by Carter, who states it thus: 'What we can expect to observe must be restricted by the conditions necessary for our presence as observers'. In short, the observers restrict the observed. The weak principle would seem not to apply to Hoyle's example, because while our presence can obviously determine the epoch t_{now}, it can surely have no influence on the structure of nuclei?

Hoyle's example can be made to comply with the weak anthropic principle, however, if one is prepared to entertain the possibility that the fundamental constants, such as α and g_s, vary throughout space or time. It would then be the case that observers would only arise in those regions of the universe where, by chance, the arrangement of nuclear resonances came out just right. Hoyle did, in fact, suggest a possible variation of this type.

Similar remarks apply to the cosmological repulsion discussed in Section 4.5. The fact that Λ is exceedingly small, due to apparently miraculous fine-tuning of Λ_q and Λ_{bare}, clearly has relevance for living organisms. A slight mismatch would produce a universe that either collapses, or explodes, catastrophically, ruling out any chances of its cognizibility. At first this fact seems merely to confirm how fortunate we are to be here. However, if Λ_q is allowed to vary, then only in relatively restricted regions of space and time will the almost exact cancellation against Λ_{bare} occur. In those regions life will be able to form, so it is no surprise that we find ourselves situated in such a region of spacetime where $\Lambda \simeq 0$. A possible reason why Λ_q might vary throughout space has been proposed by the author and Stephen Unwin.

One can extend these ideas to all the other examples of remarkable coincidences discussed in the foregoing chapters. Variations in α and g_s would produce varying admixtures of primeval hydrogen and helium, variation in the initial conditions on the big bang would produce some regions of the universe with galaxies, while others would merely possess distended clouds of gas, or black holes. Those regions that expanded initially at just the right rate remain quiescent long enough for life to form, while other, uninhabited, regions collapse or explode, or run amok with large amounts of anisotropy and inhomogeneity. And so on.

The weakness of these arguments is that there is little or no evidence for the variations in either the initial conditions or in the fundamental constants necessary for a whole range of values to exist. Strong limits on variation in time can be placed on most of the constants listed in Table 1, while variations in space would show up in the behaviour of distant galaxies, all of which seem to be remarkably similar to our own. Only variations on scales much larger than the Hubble radius would pass undetected.

An alternative scheme of thought has been established for the cases of the many extraordinary coincidences that cannot be explained by the weak anthropic principle. This involves appealing to the *strong* anthropic principle, defined by Carter as follows: 'The Universe must be such as to admit the creation of observers within it at some stage.'

Now clearly the strong anthropic principle is founded on a quite different philosophical basis from the weak principle. Indeed, it represents a radical departure from the conventional concept of scientific explanation. In essence, it claims that the universe is tailor-made for habitation, and that both the laws of physics and the initial conditions obligingly arrange themselves in such a way that living organisms are subsequently assured of existence. In this respect the strong anthropic principle is akin to the traditional religious explanation of the world: that God made the world for mankind to inhabit.

Support for the strong principle may be found in the philosophy of positivism: in essence, this demands that only that which is perceived enjoys true reality. Adopting this stance, one might argue that a universe which did not admit observers is meaningless. The only truly *real* universe is one that is perceived, so this universe is obliged to adjust its properties to whatever outrageously improbable arrangement is necessary for conscious beings to arise.

Many scientists have expressed support for the strong anthropic principle. Joseph Silk, in discussing how relation (3.14) inevitably follows in a universe that can make galaxies, remarks 'Gravitational instability and fragmentation must lead from giant clusters to galaxies to stars, and ultimately to planets and an environment suitable for the development of life. This unbroken chain is essential in any cognizable universe, and may therefore provide the key to understanding the significance of the fundamental dimensionless numbers of astrophysics and cosmology'. Notice that Silk suggests it is cognizability that will *explain* in numbers, not vice versa.

Similarly Wheeler discusses what he terms 'our participatory universe' in which the existence of an observer at some stage in its history is actually made directly responsible for the creation of that particular type of universe. For example, in discussing why the universe is so big he writes: 'What good is a universe without awareness of that universe?' He points out that unless the distance of our horizon is $\gtrsim 10^9$ light years, the universe would collapse in a time less than t_*, thereby precluding life. Wheeler concludes that the universe is so big ($\gtrsim 10^9$ light years) 'Because only so can man be here!'

Barrow reaffirms the principle that our existence is actually in some sense *responsible* for the very special structure of the universe when he writes: 'Many observations of the natural world, although remarkable *a priori*, are seen in this light as inevitable *consequences* of our own existence' (my italics).

From the strictly physical point of view it seems mysterious, to say the least, that the existence of conscious beings can actually bring about the celebrated coincidences. Clearly any direct causal connection is impossible. Special physical conditions may produce man, but man can hardly be attributed the credit for establishing his own environmental requirements.

There is, however, one area of physics in which the observer does play a central role: quantum theory. The process of measurement in quantum physics, for so long bedevilled by paradox, appears to demand the participation of the conscious observer at a fundamental level. Although the quantum observer cannot be said to actually create his own universe in the conventional sense of the word 'create', an analysis of quantum measurement theory does open the door to providing a plausible physical, as opposed to philosophical, justification for the strong anthropic principle.

5.4 The many-universes theory

The use of words 'coincidence', 'extraordinary' and 'remarkable' in the discussions of the various special relations described in this book carry the implication of *improbability*. Yet the notion of likelihood only makes sense if there is a range of alternatives from which a particular choice is made.

If a golfer tees-off at random, and happens to score a hole in one, he should think himself lucky because such an outcome is *a priori*, highly unlikely. The improbability stems from the fact that there are vastly many more locations on a golf course than the hole in the middle of the green, and that a ball driven at random will be equally likely to land at any of them. Simple statistics requires that the chances of landing in the hole are remote. Alternatively one could say that after an enormous number of random drives, only a tiny fraction will produce a hole in one. Now it could be argued that any final location of the golf ball is equally improbable. However, the point is that the hole on the green has a very special significance (at least for golfers) that is not possessed by a random patch of grass (which is why everybody claps if the ball arrives there). Similarly, the existence of life has a very special significance for us.

One could envisage a huge collection of possible universes – a world-ensemble – each varying slightly from the others so that somewhere among the ensemble would be a universe in which every conceivable value for each fundamental constant, and every conceivable initial arrangement of matter and motion, were realized to within a certain accuracy. The famous coincidences then acquire a rather more concrete status. Imagine the Creator equipped with a pin, blindly choosing one of the universes at random from among the vast collection of contenders. The chances of His picking a universe compatible with life as we know it is then exceedingly small.

It is hard to quantify the improbability of the choice of our perceived world, because although we know how to measure the relative probabilities of, say, heads or tails following the toss of a coin, we do not know how to measure probabilities between possible universes. Nevertheless, accepting the world-ensemble concept enables one to assert the general fact that our world is indeed extremely unlikely on *a priori* grounds, and that we are immensely fortunate to exist, even if we cannot assert precisely *how* fortunate.

Many people of a religious persuasion will no doubt find support from these ideas for the belief that the Creator did *not* aim the cosmic pin at random, but did so with finely computed precision, with the express purpose of selecting a universe that *would* be suitable for habitation.

Those who prefer a scientific perspective and language might turn to the subject of the quantum theory, with its inherent probabilistic structure. In principle, quantum theory ought to provide a definite probability measure for the various possible initial motions of the universe, so that one could compute, for example, how likely an outcome is the present very low degree of anisotropy given a certain quantum state for the universe.

Unfortunately, the concept of the quantum state of the whole universe is hopelessly vague, even ambiguous, in the conventional (so-called Copenhagen) interpretation of the quantum theory. The basic problem is that a quantum state can consist of a superposition of several alternative possible worlds. On measurement, one particular world – the actual world – is selected, apparently at random. The probability of the measurement producing a particular outcome can be computed, but in general any one of a range of outcomes is possible.

However, the act of measurement here must be investigated more closely. To measure a system we must have some measuring apparatus that is not itself part of the system. The act of measurement consists of temporarily coupling the apparatus to the system of interest, and allowing the system to trigger an observable change in the apparatus. When the system consists of the entire universe, the notion of a piece of external measuring apparatus is meaningless. The universe is everything that exists. On the other hand, if there is nothing left with which to measure the state of the universe, how can the universe make the transition from a superposition of many possible worlds to one, concrete, *actual* world?

This conundrum has plagued the conceptual foundations of the quantum theory for decades. Only one comprehensive resolution has ever been proposed. The basic idea is to accept the simultaneous reality of *all* the possible alternative universes. Proposed by Hugh Everett in 1957, this so-called many-universes interpretation of quantum theory provides a natural framework for the strong anthropic principle.

Before discussing the anthropic connection, it is helpful to consider as an illustration a simple scattering experiment. Suppose an electron is fired directly at a proton. The quantum wave associated with the electron diffracts from the proton and spreads outwards, like sound waves echoing in all directions from a solid object. The wave provides a measure of the probability of finding the electron at that location: where the wave disturbance is greatest there the electron is most likely to be found.

The wave scatters both left and right. Yet there is only *one* electron. As it cannot come to bits, the electron can only scatter *either* to the left *or* to the right, with certain probabilities. A measurement will reveal which is the case, but then after the measurement the wave pattern must instantly change, for if the electron is found on the right, there is no longer any probability at all that it could be on the left. The left-moving wave must then suddenly disappear.

This abrupt collapse of the wave is the essence of the quantum measurement paradox, for if the apparatus is also described by a wave, (as it must be if it is subject to quantum principles too) then even if the electron wave collapses, the apparatus wave will not – unless the apparatus is in turn measured by some further apparatus, and so on. When the entire cosmos is encompassed within this quantum description, nothing is left to collapse the wave.

In the usual interpretation of the quantum theory no real attempt is made to deal with cosmological questions. The measurement of the electron's position is regarded as projecting the world into either a state with a right-moving electron, or a left-moving electron, but not both. The state of the apparatus is left vague.

By contrast, the Everett interpretation asserts that, on measurement, the universe divides into two, with one portion containing a right-moving electron, the other a left-moving electron. Each world is equally real. Both co-exist, but do not, at least at the macroscopic level, interfere with each other. The conscious observer also splits, one copy inhabiting each world.

We must envisage every atom in every galaxy as continually engaging in this type of scattering activity, thereby splitting the world again and again into a stupendous number of near-carbon-copies of itself. The universe must therefore be likened to a tree, which branches and rebranches. Nearby branches differ little from each other, perhaps distinguished only by the arrangement of a few individual atoms. However, amid the infinite array of parallel worlds will be examples representing all possible physical universes.

This quantum-generated world-ensemble presumably contains all possible initial arrangements of matter, energy and motion. One could extend the idea and suppose that all possible values of the fundamental constants are also realized (though that was not part of the original theory). It therefore endows with reality what was formerly purely a conceptual device for discussing the improbability of our particular universe. According to the Everett theory, our very special cosmos is only one example of a limitless variety of actually existing universes.

If one accepts the Everett interpretation of quantum theory, the strong anthropic principle is no longer needed. The weak principle suffices to explain all the much-discussed coincidences, for amid the vast array of different co-existing universes, there will always be some (albeit a very small proportion) in which the numbers and conditions come out just right. Only in those universes can life form and develop. It is only those, in which the numerical relations are satisfied, that are observed. It is then no surprise that we perceive a universe to which so many very special circumstances pertain, for we have selected it from the ensemble by our very existence, just as we have selected the surface of a planet from amid a vast array of less habitable cosmic locations.

According to this viewpoint, the very special features of the universe are no longer to be regarded as extraordinary or remarkable, but as inevitable. Their apparent improbability is purely a reflection of their unrepresentative nature. The overwhelming majority of universes do not enjoy the conditions consistent with life. Only those rare ones that do are observed.

The Everett theory is not the only possible way to bestow reality on the conceptual device of a world-ensemble. Wheeler has discussed the concept of a sequential ensemble within the context of a recontracting cosmological model. Such a universe, it will be recalled, expands from an initially singular condition to a maximum volume, and then collapses back to total obliteration at a final spacetime singularity. If the singularity is taken seriously, it represents the complete breakdown of all known physics. Wheeler exploits this opportunity by proposing that some sort of universe in fact survives its encounter with the singularity, but emerges 'reprocessed', with new values for the fundamental constants, a new pattern of motion, and perhaps even new laws of physics.

The formerly collapsing universe thus 'bounces' out again, revitalized, into a new cycle of expansion and contraction, to be followed by yet another, and so on, *ad infinitum*. In each cycle the structure of the universe is different. If the reprocessing is performed at random, then eventually, purely by chance, the numbers and organization will concur felicitously and the various numerical relations will appear. These cycles will permit the development of cosmologists, who will write books about the extraordinary degree to which nature has conspired to arrange its affairs for the benefit of living beings.

Barry Collins and Stephen Hawking, appealing to the notion of an ensemble of worlds, addressed the question: 'Why is the universe isotropic?' In Section 4.3 it was argued that an anisotropic universe might produce vast quantities of heat, which would prevent the formation of galaxies by exerting strong radiation pressure. Obviously such circumstances would not favour life as we know it. Collins and Hawking approached the issue from a different angle. They demonstrated that, in general, the universe ought to become more and more anisotropic as it expands. However, if the expansion rate is *exactly* matched to the gravitating power so that $k = 0$ (that is, condition (4.14) is satisfied), then it will remain isotropic. They argue that only in a universe close to the $k = 0$ condition can long-lived galaxies form: when k is appreciably less than zero, the expansion is too vigorous to allow gravitational clumping, and when k is appreciably greater than zero, the universe recollapses rather rapidly. Collins and Hawking conclude that only in a universe with $k \simeq 0$, hence one which can remain isotropic for a long time, will life form. Amid the ensemble of co-existing universes, almost all of which do not have $k \simeq 0$ and which are highly anisotropic, we have selected an isotropic one in which galaxies provide suitable conditions for our existence. To their question of why the universe is so isotropic, Collins and Hawking answer: 'Because we exist'.

The world-ensemble concept can be criticized on a number of grounds, some philosophical, some physical. It may appear unattractive that nature should indulge in such profligate duplication. Can we really believe in limitless numbers of universes, created but never observed, serving no purpose except to ensure that, somewhere among the vast array of wasted worlds, will be the occasional cognizable accident? To explain the coincidences by invoking an infinity of useless universes seems like carrying excess baggage to the extreme. Yet it must be conceded that the alternatives – a universe deliberately created for habitation, or one in which the very special structure is simply regarded as a pure miracle – are also open to philosophical challenge.

The anthropic principle based on a randomly arranged world-ensemble has also been criticized on physical and mathematical grounds. The problem goes back to an old idea of Ludwig Boltzmann that the present high degree of cosmic organization is the result of a stupendously rare statistical fluctuation from a far more probable condition of featureless disorder, and that the only reason that we are privileged to witness this exceedingly unlikely occurrence is that our very existence depends upon the conditions which alone can be established by that remarkable fluctuation. This is, of course, an early statement of the weak anthropic principle.

The fluctuation to which Boltzmann refers is simply a cosmic-sized version of the sort of fluctuations that produce Brownian motion among tiny particles suspended in a fluid. Given the random cavorting of all atoms, wholesale cooperation of large numbers of atoms will conspire, after an unbelievably long duration, to produce order spontaneously out of chaos – like the chimpanzee who plays Beethoven accidentally by unending tinkering on the piano. A simple example concerns a box of gas: wait long enough and, purely by chance, all the gas molecules will rush simultaneously to one end of the box. Though the wait is enormously long, in an infinite time everything is possible.

Now an essential feature of order achieved spontaneously through random fluctuations is that it is overwhelmingly more likely for a small quantity of order to develop than a large amount. The chimpanzee is far, far more likely to hit upon the first bar of 'three blind mice' than to achieve a sonata.

By the same token, a fluctuation which produces, say, one galaxy, is overwhelmingly more likely than one which produces billions of them. Yet one galaxy is surely enough to produce conscious observers. Why then, do we continue to see order the farther we look into space?

The same criticism may be directed against the anthropic principle as applied to the world-ensemble. There will be vastly more universes in which the arrangements are precisely tuned to produce a single galaxy, than there will be multi-galactic universes. A typical observer will therefore, on the face of it, be far more likely to find himself in a one-galaxy cosmos. The ubiquity of observed galaxies in *our* universe is therefore a mystery.

This challenge, which has been deployed by Roger Penrose against the anthropic principle, ignores any coupling which may exist between local and global structure. It may be that there is, in fact, a link between the formation of galaxies and the large scale arrangement of the cosmos. A link of this sort is provided by the so-called Mach principle, which attributes the origin of inertia to interactions with distant galaxies. R. Dicke and P.J.E. Peebles, in commenting that the universe seems 'over designed' for the modest purpose of a few conscious individuals, remark that a single-galaxy cosmos might be ruled out by Mach's principle.

Using the world-ensemble hypothesis combined with the weak anthropic principle, it is possible to discover plausible arguments that pin down the values of almost all of the fundamental parameters which, as discussed in Chapter 2, more or less determine the structure of the physical world. It is possible that more detailed analyses could constrain other features such as the dimensionality of space and time, the numbers of quarks and leptons, the numbers of fundamental forces, and so on.

Of course, such arguments are no substitute for a proper physical theory. It is hard to see, for example, how the anthropic principle can ever be used to make a testable prediction, because any physical theory that is inconsistent with our existence is manifestly incorrect anyway. Moreover, in the absence of knowledge about extra-terrestrial life, we have only rather general arguments about the physical prerequisites for biology. Perhaps life can form under a much wider variety of conditions than hitherto assumed.

It might well be that future developments will provide for some of the numerical coincidences discussed in the foregoing chapters explanations based on fundamental physics rather than biology. The ratio of the strengths of the forces, for example, could emerge from a forthcoming super-unified theory of GUTs with gravity. In that case the mysterious 10^{40} will be derivable from mathematics. Similar reasons might be discovered for the homogeneity and isotropy of the universe. Hitherto unsuspected processes occurring in the little-understood primeval universe could have forced the cosmic motion into its otherwise unexpectedly symmetric behaviour.

Given such future success in providing basic physical reasons for the apparently accidental arrangement of the world, the anthropic principle would lose any explanatory power. Nevertheless, it would be no less remarkable that basic physics had been found to be organized in a fashion so propitious for life. Whether the laws of nature can force the coincidences on the universe or not, the fact that these relations are necessary for our existence is surely one of the most fascinating discoveries of modern science.

BIBLIOGRAPHY

Chapter 1

The Forces of Nature by P.C.W. Davies (Cambridge University Press, 1979) provides an introduction to quantum theory, quantum fields and modern particle physics at about the same level as this book.

For an introduction to the subject of modern cosmology, with special emphasis on the early stages of the universe, the reader might like to try *The First Three Minutes* by Steven Weinberg (Basic Books, New York, 1977) or *The Runaway Universe* by Paul Davies (J.M. Dent, London, 1978; Harper & Row, New York, 1978).

More advanced texts are *Principles of Cosmology and Gravitation* by Michael Berry (Cambridge University Press, 1976) and *Modern Cosmology* by D.W. Sciama (Cambridge University Press, 1971).

A good all-round introduction to astrophysics and cosmology is *The State of the Universe* (ed. Geoffrey T. Bath, Clarendon Press, Oxford, 1980). In particular the articles by Rees on galaxy formation, Tayler on nucleosynthesis in the big bang and inside stars, and Sciama on basic cosmology, are especially relevant to the discussion in this book.

Chapter 2

This chapter is based on an excellent review article, 'The Anthropic Principle and the Structure of the Physical World' by B.J. Carr & M.J. Rees, *Nature* **278**, 605 (1979) wherein further references can be found.

The issue of the 'constancy of the constants' has been explored by F.J. Dyson in *Aspects of Quantum Theory* (eds. A. Salam & E.P. Wigner, Cambridge University Press, 1972); see also the discussion in *Gravitation* by C.W. Misner, K.S. Thorne & J.A. Wheeler (Freeman, San Francisco, 1973), Chapter 38.

Chapter 3

Full details of the thermodynamic properties of the cosmological material in the early stages of the universe, especially the issue of neutrino decoupling and nucleosynthesis, are given in *Gravitation and Cosmology: Principles and*

131

Applications of the General Theory of Relativity by Steven Weinberg (Wiley, New York, 1972). The treatment is rather advanced; for an elementary and more descriptive account, the reader may consult the cosmology part of *Black Holes, Gravitational Waves and Cosmology* by M.J. Rees, R. Ruffini & J.A. Wheeler (Gordon & Breach, New York, 1974), and *Modern Cosmology* by D.W. Sciama (Cambridge University Press, 1971).

Freeman Dyson's remarks on the di-proton may be found in his article in *Scientific American*, **225**, 25 (September 1971).

Brandon Carter's analysis of stellar structure treated in Section 3.3 is summarized in his article in *Confrontation of Cosmological Theories with Obervation* (ed. M.S. Longair, Reidel, Dordrecht, 1974).

The analysis of galactic structure given in Section 3.4 is based upon the work of J. Silk: see *Nature* **265**, 710 (1977).

Chapter 4

There is a huge literature on the famous large-number coincidences. The original references seem to be A.S. Eddington, *Proc. Cam. Phil. Soc.* **27**, 15 (1931); *Relativity Theory of Protons and Electrons* (Cambridge University Press, 1936) and P.A.M. Dirac, *Nature* **139**, 323 (1937); *Proc. R. Soc.* **165A**, 199 (1938).

Later elaborations include the work of P. Jordan, *Schwerkraft und Weltall* (Viewig & Sohn, Braunschweig, 1955); R.H. Dicke, *The Theoretical Significance of Experimental Relativity* (Gordon & Breach, New York, 1964); S. Hayakawa, 'Atomism and Cosmology', *Prog. Theor. Phys., Supplement*, Yukawa 30th anniversary issue, 532 (1965), and B. Carter, 'Large numbers in Astrophysics and Cosmology', Institute of Theoretical Astronomy (Cambridge preprint, 1968). See also, 'The "Large Numbers": Coincidence or Consequence' in *Black Holes, Gravitational Waves and Cosmology* by M.J. Rees, R. Ruffini & J.A. Wheeler (Gordon & Breach, New York, 1974).

A review of the unification of the fundamental forces, and implications for proton decay has been given by Steven Weinberg in *Scientific American* **244**, 52 (June 1981).

For a possible perspective on the relation (4.12) and a connection with Mach's principle, see *The Unity of the Universe* by D.W. Sciama (Anchor, New York, 1961).

For a discussion of cosmic dynamics, a good, elementary treatment is given in H. Bondi's classic book *Cosmology* (Cambridge University Press, second edition, 1961), and in Michael Berry's *Principles of Cosmology and Gravitation* (Cambridge University Press, 1976). Slightly more advanced, but with a transparent explanation of the topic of horizons, is *Essential Relativity* by

W. Rindler (second edition, Springer-Verlag, New York, 1977). This book also treats the subject of the cosmological constant, Λ. *The Large Scale Structure of the Universe*, by P.J.E. Peebles (Princeton University Press, 1980), which also treats galaxy formation and anisotropy, is a good, up-to-date, advanced text.

The fine-tuning of ρ and ρ_{crit} discussed in Section 4.2, is sometimes known as 'the flatness problem', while the lack of communication between far flung galaxies is called 'the horizon problem'. Both are analyzed in depth (and a possible resolution suggested) in a paper by A. Guth: *Phys. Rev.* **D 23**, 347 (1981). This was the origin of the so-called inflationary universe scenario, since developed in great detail. For a review see J.D. Barrow & M.S. Turner, *Nature* **298**, 801 (1982).

These topics are also dealt with by R.H. Dicke & P.J.E. Peebles, 'The Big Bang Cosmology – Enigmas and Nostrums', in *General Relativity: an Einstein Centenary Survey* (eds. S.W. Hawking & W. Israel, Cambridge University Press, 1979).

The dissipation of primeval turbulence is studied in detail in a paper by J.D. Barrow & R.A. Matzner, *Mon. Not. R. Astr. Soc.* **181**, 719 (1977).

Speculations on the origin of the entropy per baryon, S, are widespread. Section 4.4 referred to the work of M.J. Rees, *Nature* **275**, 35 (1978). Within the context of baryon non-conservation, D.V. Nanopoulos has studied S, in an anthropic spirit, in *Physics Lett.* **91B**, 67 (1980). This paper also refers to the original calculations in which the quoted values for S are calculated.

Three important papers on the subject of the cosmological constant and its relation to quantum vacuum effects are D.A. Kirzhnits & A.D. Linde, *Ann. Phys.* (NY) **101**, 195 (1976); V. Canuto & J.F. Lee, *Physics Lett.* **72B**, 281 (1977); S. Coleman & F. DeLuccia, *Phys. Rev.* **D 21**, 3305 (1980).

Chapter 5

Although the words 'anthropic principle' are not new, they appear to have been used first in a modern context by Brandon Carter. His article in *Confrontation of Cosmological Theories with Observation* (ed. M.S. Longair, Reidel, Dordrecht, 1974) is the first to state clearly the weak and strong versions, and their relationship to astrophysical processes.

The most complete discussion of the subject, including a great deal of historical material, has been given by J.D. Barrow & F.J. Tipler, *The Anthropic Principle* (Oxford University Press, 1982). The philosophical ramifications have been studied by John Leslie in his article 'Anthropic Principle, World Ensemble and Design', to appear in the *American Philosophical Quarterly*.

Robert Dicke's explanation of one of the large-number coincidences, discussed in Section 5.2, appears in a short paper in *Nature*, **192**, 440 (1961).

Much of this early work, together with more recent ideas, is reviewed by B.J. Carr & M.J. Rees in *Nature* **278**, 605 (1979). Another wide-ranging review of 'anthropic coincidences' is given by I.L. Rozental in *Soviet Physics Usp.* **23**, 296 (1980).

Dyson's analysis of the survival prospects for intelligent organisms in a decaying cosmos are given in *Rev. Mod. Phys.* **51**, 447 (1979).

The crucial importance of the nuclear resonances for the synthesis of carbon in stars was pointed out by Fred Hoyle in *Astrophys. J. Supplement* **1**, 121 (1954). (See also *Galaxies, Nuclei and Quasars*, Harper & Row, New York, 1964.) The quotation at the end of Section 5.3 refers to an unpublished University of Cardiff preprint entitled 'The Universe: Some Past and Present Reflections' by Fred Hoyle.

The possibility of a position-dependent Λ has been discussed by the author and Stephen Unwin in *Proc. R. Soc.* **A 377**, 147 (1981).

Another model of an inhomogeneous universe has been considered by G.F.R. Ellis, *Gen. Relativ. Gravit.* **9**, 87 (1978).

John Wheeler's concept of the cyclic 'reprocessing' universe is described in Chapter 44 of *Gravitation* by C.W. Misner, K.S. Thorne & J.A. Wheeler (Freeman, San Francisco, 1973). The central role of the observer (the so-called 'participatory universe') is explained in a quantum context in his article in *Some Strangeness in the Proportion: A Centennial Symposium to Celebrate the Achievements of Albert Einstein* (ed. H. Woolf, Addison-Wesley, Reading, Mass., 1980).

The many-universes interpretation of quantum mechanics originated with H. Everett, *Rev. Mod. Phys.* **29**, 454 (1957), and was extensively developed by B.S. DeWitt & N. Graham in *The Many-Worlds Interpretation of Quantum Mechanics* (Princeton University Press, 1973).

The work of C.B. Collins & S.W. Hawking on the anthropic explanation of cosmic isotropy may be found in *Astrophys. J.* **180**, 317 (1973).